T0141164

About Island Press

Since 1984, the nonprofit Island Press has been stimulating, shaping, and communicating the ideas that are essential for solving environmental problems worldwide. With more than 800 titles in print and some 40 new releases each year, we are the nation's leading publisher on environmental issues. We identify innovative thinkers and emerging trends in the environmental field. We work with world-renowned experts and authors to develop cross-disciplinary solutions to environmental challenges.

Island Press designs and implements coordinated book publication campaigns in order to communicate our critical messages in print, in person, and online using the latest technologies, programs, and the media. Our goal: to reach targeted audiences—scientists, policymakers, environmental advocates, the media, and concerned citizens—who can and will take action to protect the plants and animals that enrich our world, the ecosystems we need to survive, the water we drink, and the air we breathe.

Island Press gratefully acknowledges the support of its work by the Agua Fund, Inc., Annenberg Foundation, The Christensen Fund, The Nathan Cummings Foundation, The Geraldine R. Dodge Foundation, Doris Duke Charitable Foundation, The Educational Foundation of America, Betsy and Jesse Fink Foundation, The William and Flora Hewlett Foundation, The Kendeda Fund, The Andrew W. Mellon Foundation, The Curtis and Edith Munson Foundation, Oak Foundation, The Overbrook Foundation, the David and Lucile Packard Foundation, The Summit Fund of Washington, Trust for Architectural Easements, Wallace Global Fund, The Winslow Foundation, and other generous donors.

The opinions expressed in this book are those of the author(s) and do not necessarily reflect the views of our donors.

Floodplain Management

Floodplain Management

A New Approach for a New Era

Bob Freitag, Susan Bolton,
Frank Westerlund, and J. L. S. Clark

ISLANDPRESS
Washington | Covelo | London

ISLAND PRESS is a trademark of the Center for
Resource Economics.

Floodplain management : a new approach for a new
era / Bob Freitag . . . [et al.].
    p. cm.
  Includes bibliographical references and index.
  ISBN-13: 978-1-59726-634-5 (cloth : alk. paper)
  ISBN-13: 978-1-59726-635-2 (pbk. : alk. paper)
  ISBN-10: 1-59726-635-3 (pbk. : alk. paper) 1.
Floodplain management—United States. 2. Flood-
plain management—United States—Case studies. I.
Freitag, Bob.
  TC423.F635 2009
  627'.4—dc22   2009009800

Printed on recycled, acid-free paper

Manufactured in the United States of America
10 9 8 7 6 5 4 3 2 1

Keywords: flood, river, natural floodplain management,
wetland, swale, levee, Corps of Engineers (USACE),
Natural Resources Conservation Service (NRCS),
Federal Emergency Management Agency (FEMA),
National Flood Insurance Program (NFIP), landscape
design, climate change, natural hazards, resilience,
habitat conservation, Mississippi River, Hurricane
Katrina

Contents

Acknowledgments

Wayne Blanchard, FEMA's Higher Education Project Manager, was largely the inspiration for this book. In 2005, he asked a group of us to create a graduate-level floodplain management course. Susan Bolton, Bob Freitag, and Frank Westerlund, along with Don Reichmuth, Elliot Mittler, and the late Rod Emmer, developed that course. (It is currently on the FEMA Higher Education Project Web site and this book can be used as its text.) The FEMA course was based on an earlier integrated floodplain management course developed for the Washington State Department of Ecology under the direction of Tim D'Acci. Christina Hill and Bill Leon worked with Bolton and Freitag on the Department of Ecology project. Many of the concepts in this book were based on sessions developed for those earlier courses and we are deeply indebted for the original work done by our course coauthors.

We held two sets of workshops at which we tested the content of this book. Thanks to Mark Darienzo and Christine Shirley, both of the Oregon Department of Land Conservation and Development, and to Andre LeDuc, the Program Director of the Oregon Natural Hazards Workgroup at Oregon State University, for their help in delivering Oregon workshops in Bend and Salem.

The book went through many stages. Concepts were drafted, reviewed, debated, written, and rewritten. Special appreciation goes to friends and colleagues who helped enormously along the way: Bob Aldrich, David Carlton, David Conrad, Daniel Friedman, Hans Hunger, Ryan Ike, Irina Kukina, Larry Kunzler, Jon Kusler, Larry Larson, Luke Meyers, Ann Patton, Mark Riebau, Robert Schneider, Norman Skjelbreia, Harold Smelt, Dan Sokol, Chuck Steele, Takahiro Tanaka, Ed Thomas, Ronald Throupe, Fritz Wagner, Coburn Ward, French Wetmore, and Chuck Wolfe. Any mistakes, of course, remain the responsibility of the authors.

We also thank the graduate students who took our floodplain management courses and participated in experiments with both content and approach. Of special significance is the work done by Sarah Hawkins, who wrote her thesis on the National Flood Insurance Program and gave us opportunities to view this program from different perspectives. We also thank the students involved in discussions with us who broadened, altered, and clarified our understanding of floodplain management, including: Lauren Acheson, Jeffrey Aken, Nick Arcos, Wendy Buffett, Rebecca Buttitta, Rebecca Chaney, Vivian Chang, Kelly Clark, Glenn Coil, Richard Collins, Patrick Dagon, James Dewar, Amanda Engstfeld, Marion Gonzalez, Sarah Hawkins, Megan Horst, Christine Howard, Zuzana Hrasko-Johnson, Heidi Kandathil, Patrick Keys, Khalid Khan, Kuei-Hsien Liao, Tzu-Yu Lin, Beth Martin, Kara E. Martin, Derek Mason, Erick McConaghy, Kristen Meyers, Rachel Minnery, Andrew Redman, James Rufo Hill, Christopher Scott, Dana Spindler, David Sullivan, Peter Sullivan, Fred Swenson, Lucy Walsh, and Scott Williamson.

Author Bob Freitag was on the Association of State Floodplain Managers (ASFPM) Certification Board (CBOR), and later the ASFPM board of directors. This gave him access to some of the most talented floodplain management practitioners in the nation. Thanks go to the CBOR members, including Michael Borengasser, Collis Brown, Diane Calhoun, Cindy Crecelius, Tom Hirt, John Ivey, Anita Larson, Rhonda Montgomery, Mike Parker, George Riedel, Cleighton Smith, Ann Yakimovicz, and Ken Zwickl.

We very much appreciate Island Press for agreeing with us on the importance of our philosophy of flood management and especially thank Senior Editor Heather Boyer's enthusiastic shepherding of the book (and us) through their process. We quite literally could not have done it without her.

Very special thanks go from each of the authors to their families, to Wendy Freitag and Fereshteh Westerlund, whose patience and support are appreciated but sometimes taken for granted.

And finally we must thank Bob Martin, an Oregon rancher and retired Natural Resources Conservation Service employee, for his undaunted support of the potential of good, old-fashioned conservation practices. While touring the Buck Hollow watershed, this exceptional conservationist outlined a practical approach through the creation of hundreds of small inexpensive retention/detention structures, with associated downstream benefits. This strategy is proving increasingly important as our climate warms and our land becomes covered with structures.

Chapter 1

Floods Are Not the Problem

Rivers will do what rivers do. Historic flooding of the Mississippi River shows that our approach to flood control hasn't worked and can have effects far beyond the limited area of a floodplain. The consensus of scientists around the world is that we are in a period of rapid global climate change, which makes working with rivers—instead of against them—increasingly important. Some places will get more intense rainstorms, while others may get less frequent summer rains. Still other areas may get winter storms with less snow and more rain, producing immediate runoff instead of storing water for spring. These and other changes may increase flooding, and, in some cases, may also increase drought.

In many cases, flood management practices are based on dam or levee projects that do not incorporate all we now understand about river processes. They try to control the river. Many years of experience with dam and levee systems have shown their limitations. Though dams and levees may be necessary in some cases, more often a larger suite of tools is available. We suggest instead that a better solution is to work with the natural tendencies of the river: *retreat* from the floodplain by preventing future development and sometimes even removing existing structures; *accommodate* the effects of floodwaters with building practices; and *protect* assets with nonstructural measures if possible, and large structural projects only if absolutely necessary.

To help decide on the best (cheapest, longest lasting, most beneficial, and so on) project choice to control destructive floods or enhance our water resources, we should answer six questions:

1. What values or assets do you want to protect or enhance?
2. What are the apparent risks or opportunities for enhancement?

3. What is the range of risk-reduction or opportunity-enhancement strategies available?
4. How well does each strategy reduce the risk or enhance the resource?
5. What other risks or benefits does each strategy introduce?
6. Are the costs imposed by each strategy too high?

Only after answering all six questions will we find the optimal strategy.

Lessons from the Mighty Mississippi

The Mississippi watershed stretches from Canada to the Gulf of Mexico between the Rockies and the Appalachians. The Mississippi River and its tributaries drain all or part of thirty-one states and two Canadian provinces over 1 million square miles. Major tributaries include the Minnesota, Wisconsin, St. Croix, Iowa, Skunk, Des Moines, Illinois, Ohio, Missouri, and Arkansas rivers.

As early as the 1800s, Mississippi River flooding created a problem for adjacent cities and farms, as well as for the river traffic that transported a good deal of American commerce. Many engineers of this time were educated at West Point and assigned to the U.S. Army Corps of Engineers (called USACE or the Corps). For a variety of reasons, the Corps was given the job of "fixing" the floods along the Mississippi. Levees were the weapon of choice. In the short term, the levees protected productive floodplains from being retaken by the river and actually created valuable land. By 1858, more than a thousand miles of levees had been built. These produced the typical feedback loop: levees gave floodwaters a narrower reach to flow through, making floods higher, requiring even higher levees.

The USACE continues to manage flood control for the Mississippi River. Many policy decisions were made when the watershed and our explanation of river processes were very different, and are still in effect today. The natural alluvial valley of the Mississippi is a wide swath of forest and grassland that allowed the river to roam across it, producing constantly shifting meanders. Frequent flooding left silt on the land, which provided rich farmland for early settlers. This very activity of the river—its wandering, which left such rich earth—is exactly the problem it posed for cities along the river. Urban areas need rivers to stay in a fixed location. The levees-only policy was based on a fundamentally wrong tenet: that by confining the river within levees the erosive power of the water would be redirected to the bottom of the channel. Using levees was supposed to force the river to scour its channel, making it deeper and allowing it to carry high flows without flooding. This scouring and deepening never happened.

Today, little of the Mississippi valley is in its natural condition. Forests made way for farms and cities and even much of the floodplain is urban, channeling rain to the river ever faster. This increases the height of floodwaters and requires higher levees. Though there are increasing numbers of projects that enhance outlets and impoundment to contain these high flows, most of the focus remains on levees. The result has been catastrophic flooding.

1926–1927

By 1926 the USACE had pursued a levees-only policy on the Mississippi for several decades, going so far as to seal off many natural outlets. The effect should have been clear: through the years, similar amounts of water in the river produced higher and higher floods. Still the policy continued, ultimately leading to a catastrophic flood of the lower Mississippi. Places in the upper watershed like Helena, Montana, saw intense snowfall in the winter of 1926–1927. Pittsburgh, Pennsylvania, and Cincinnati, Ohio, flooded in January 1927. March snow fell from Colorado to Tennessee. For months, record precipitation fell on the basin, ultimately reaching the river. In response the river ran high, flooding tributaries and the main channel.

Upper tributaries flooded first, starting in September 1926. Officials knew that water would eventually flow through the lower Mississippi, where 800 miles of levees were the only defense against the stream. The Corps inspected and reinforced weak spots, but the levees did not hold.

The first to break were those on tributaries, built and maintained by state governments or private contractors. The Corps continued to insist that no levee built to federal standards had ever broken. On April 16, 1927, that changed.

Over the next weeks, the Mississippi flooded 27,000 square miles over seven states, home to nearly 1 million people. In some spots, the river ran more than sixty miles wide. The last floodwaters did not recede until August. For nearly a year, one part or another of the Mississippi watershed had languished under floodwaters.

The official death toll of 246 certainly underestimates the true number of casualties. Additionally, more than 130,000 homes were lost and

700,000 people were displaced. In today's dollars, estimated property damage totaled $5 billion.

The 1927 flood changed river policy, but it had other long-lasting effects on American society.

- Herbert Hoover's substantial flood relief efforts propelled him into the White House.
- The lack of preparation and the generally reprehensible flood relief efforts were a factor in Huey Long becoming governor in Louisiana, shifting political power away from the elites who had ruled New Orleans and its environs for generations.
- The destroyed economy of the delta region and deplorable treatment of many African Americans during the flood added to the Great Migration of African Americans out of the South and to northern cities. As many as 50 percent of Mississippi delta country African Americans moved north within a few years of the flood.
- Before the great flood, government did not provide relief services to victims of floods or other catastrophic events. President Coolidge refused to get involved, voicing the traditional opinion that government aid demeaned its recipients and that communities should take care of their own. The enormous scale of the 1927 catastrophe required a federal approach. From that time on, though the programs have changed considerably, the federal government has been expected to take care of its most unfortunate citizens in a catastrophe.

Flood control policy changed unalterably. The vastness of destruction discredited the idea of using only levees. The Corps had to come up with a new way to keep the Mississippi from flooding and destroying the inhabitants along its banks. Under the Flood Control Act of 1928, levees remained key but were supplemented with meander cutoffs, flood outlets, reservoirs upstream and on tributaries, and other measures. By 1936 the Mississippi River had twenty-nine locks and dams, hundreds of runoff channels, and a thousand miles of levees.

At the same time, development continued within the watershed, changing forestland to farmland and farmland to cities, altering the basic ecology and hydrology of the watershed.

1993

In 1927, the lower Mississippi took the brunt of flooding. In 1993, water overwhelmed the upper watershed. Significant flood control changes made after the 1927 and subsequent floods were supposed to prevent catastrophic flooding. They did not. For example, the Corps shortened the river by more than 150 miles by creating cutoffs that eliminated a series of sharp curves. These initially lowered flood heights 15 feet but the river has regained one third of these cutoffs.

In 1993 precipitation and runoff were intense in the western part of the Mississippi's basin along the Minnesota, Iowa, Des Moines, Kansas, and Missouri rivers. Saturated soils in the Midwest contributed to record runoffs for these and other streams. At St. Louis the river flowed more than 50 feet above flood stage for more than three months. Nearly 150 major rivers and tributaries were at flood stage at the same time.

In the summer of 1993, floodwaters killed fifty-two people, damaged more than 1,000 levees (only 20 percent of federal levees were damaged, as opposed to 80 percent of local levees), destroyed 50,000 homes, and forced the evacuation of 70,000 people. Flooding affected more than 27,000 square miles. At least seventy-five towns were completely under floodwater. The dead zone in the Gulf of Mexico at the mouth of the Mississippi, in which low oxygen levels cannot support life, expanded. Total flood damage estimates ran as high as $20 billion. The USACE estimated that flood-control structures prevented an additional $19 billion in damage.

After the flood, more than 300 homes were moved to safer ground and 12,000 homes were bought and razed at a cost of more than $150 million. The empty lands were then dedicated to parks and wildlife habitat that could be used as temporary reservoirs in future floods. Levees were rebuilt, sometimes to higher and better standards. At least four entire towns considered moving to higher ground.

2008

Many analysts looked at the Mississippi watershed and levees after the 1993 floods. Human development has changed the natural characteristics of the basin. Most of the wetlands have been drained, cities cover much of the floodplain, and vast swaths of the basin used to be grassland with deep roots that would draw in water. Now agricultural land has been drained and tiled to keep it from flooding. All this means more water runs more quickly into rivers throughout the basin.

The main stem of the Mississippi River and many of its tributaries are narrowly confined by levees. However, these are not part of a coordinated system. The USACE builds levees, state and local governments build levees, and private interests build levees. They are built to different standards, are maintained by different entities, and each levee is engineered independently of the others. No person or agency looks at the levees as a system, researching what effects new levees have on the river and how they may increase the pressure on existing levees, mandating and enforcing a maintenance schedule, or looking at the total effect of the levees as a whole on the dynamics of the watershed. Calls for such oversight after the 1993 floods went unanswered.

In June 2008, another massive flood overwhelmed the upper Mississippi basin, stunning many who believed that such an event could not happen after the lessons of 1993. Some called it the second 500-year flood in fifteen years, highlighting the problem of that terminology. A 500-year flood happens, on average, every 500 years. It's the "on average" part that causes confusion. While some researchers called them 500-year floods, others thought the 2008 flood could be a less frequent event, possibly a 200-year flood. It is possible that one or both of those floods could be 500-year floods if the watershed were in its natural state. However, with the massive amount of urban development (that continues to increase) along the Mississippi, it is unlikely that either flood would happen, on average, only every 500 years. We are very likely to see similar flooding in the twenty-first century.

The $2 billion of damage was small only compared to the catastrophe of 1993. The destroyed homes, flooded fields, and shutdown of the river to barge traffic were expensive both in terms of money and of human endeavors. As if further proof were needed, it became clear that no integrated flood control system exists for the Mississippi watershed.

Two divergent trends helped shape the amount of damage in this flood. On one hand, buildings were removed from tens of thousands of acres after 1993, leaving land that now can be flooded. That action substantially lowered the damage in 2008. On the other hand, people continued to view new and rebuilt levees as absolute protection, and thousands of acres of floodplain behind the levees became the site of new homes and businesses. When levees have overtopped or failed, these places have flooded. Even with the inadequacies of levees repeatedly demonstrated, the allure of rich, flat land close to the river and existing urban areas is too strong to stop development.

Fundamental Problems

Mississippi River floods show that our approach to managing floods and floodwaters is fundamentally wrong.

In our industrial society, we're used to "overcoming" nature with science and technology. Don't want dirt on your clothes? Here's a new detergent. Have to travel 20 miles to get to work? Here's a bigger/smaller/safer/more luxurious car.

But we don't really control nature, and an everlasting truth is rivers will do what rivers do. That's not to say that we have to live with the whims of floods, meandering channels, and droughts. These things can be ameliorated, but if not done correctly there will be unintended consequences—some of them quite serious.

There are good reasons for people to live in and use floodplains, but their use must be regulated by society. Floodplains are an integral part of life and need to be protected, if not for themselves then for the enhancement of humanity. The only way to conserve floodplains for the long term is to understand how they work, and to work with—not against—their natural processes.

In this book, we do not use the phrase "natural and beneficial processes" even though this is a long-standing expression in watershed management. The reason is simple. These processes are integral to effective management, not a separate and discrete module of it. Our philosophy is simple: natural processes are the starting point of any successful project and they must be considered and built upon at every step, not considered as an addendum after making plans.

Rivers convey, create, and conserve a number of values and assets. Some are natural, others are man-made. A couple might want to protect their streamside home, a conservation group might want to enhance riverine natural beauty and ecological benefits, and a city might want to use the river for a community water supply.

Each use affects the others and is affected by the others. The Columbia River, for example, provides scenic beauty that attracts tourists, an important part of the economy for the Oregon, Washington, and British Columbia communities along the river. It provides a spiritual element to Native American peoples in its watershed, as well as to others who feel a connection with nature. In addition, its water provides irrigation for farms and ranches, hydroelectric power for residents and businesses as far away as California, transportation for crops and manufactured goods, and habitat for endangered salmon and myriad other wildlife among many other riverine benefits.

The essential problem with floodplain management is that there are competing interests at competing scales. The "best" management for some activities may curtail or destroy other activities. In the parched areas of the arid western United States, the battles continue for the same inadequate supply of water. Is it more important to use it for agriculture? For homes? For industry? For fish and wildlife? No answer will please everyone because there is simply too little water for too many mutually exclusive uses. Even after making a decision, the system must be periodically reviewed to see if yesterday's priorities—and available water—are the same as today's.

We already know that changing land cover can change the behavior of the watershed. Forested land in its natural state may be so dense that only a small fraction of rain ends up in a stream. Plants or soils take up the rest. Cutting down forests and using the land for agriculture allows more water to run across the land into streams. An urban area, with impervious surfaces such as roads and roofs, creates even greater and more rapid runoff.

How climate change will affect individual areas is still unknown. In some areas, rain may fall in different patterns, with a higher concentration in winter months. In others, warmer temperatures will mean less snow—but more rain—in winter. That could lead to the simultaneous problems of floods in winter and lack of water in summer.

Climate models have produced forecasts and research continues to improve their accuracy. For example, in future years there may be more rain that falls in the Mississippi watershed, increasing the number and/or height of floods. According to one study, average yearly rainfall in St. Louis will increase by 21 percent within the next thirty years—increasing the amount of water in the Mississippi River by just over 50 percent.

This increased uncertainty about precipitation and other weather will further stress our rivers and watersheds. In addition, rising sea levels will force rivers to adjust their own levels, creating another uncertainty in the future behavior of our streams.

Our society is much more focused on mopping up disasters than preventing them. A study by economists Andrew Healy and Neil Malhotra shows that voters reward government officeholders who provide disaster relief, but not those who invest in disaster prevention. This disincentive has left us with underfunded disaster prevention programs. According to their study, a dollar spent on prevention is more than ten times more valuable than a dollar spent on relief. Another study, done by the National Institute of Building Sciences, looked at a sample of hazard mitigation grants awarded by FEMA (the Federal Emergency Management Agency). Their cost-benefit analysis found that every $1 spent on mitigation saved society an average of $4.

We propose a method that acknowledges that financial lesson, helps prevent future flood losses more cost-efficiently, and at the same time preserves the environment.

Book Organization

Each chapter focuses on a different step of the decision-making process we suggest when contemplating changes to river flows. At the end of each chapter, there is an example of how our six questions can be used as decision-making tools.

- Chapter 1 introduces the importance of successful floodplain management and the six questions to ask before beginning new projects involving rivers.
- Chapter 2 discusses important concepts in watershed analysis.
- Chapter 3 gives definitions that will be used throughout the book and a framework for future discussions.
- Chapter 4 is one of the most important chapters of the book. It describes some of the physical and biological processes that must be taken into account when planning for flood mitigation.
- Chapter 5 describes how human use of riverine resources has changed, and continues to change, over time.
- Chapter 6 describes structural and nonstructural options, along with their strengths and weaknesses.
- Chapter 7 looks at the suite of tools available for managing the changes brought about by flooding.
- Chapter 8 delineates flood management strategies appropriate for a variety of stakeholders, pulling together information from previous chapters.
- Chapter 9 gives several systems for evaluating the different strategies that you are considering and asks you to make a final decision.
- Chapter 10 asks you to take action.
- Appendix A gives detail on the National Flood Insurance Program, a critical piece of America's flood management practice.
- Appendix B presents a graphic tool kit that can be used to sketch (or use a geographic information system for) current or future watershed conditions.

Six Questions

We'll show you how to integrate nature and society to protect and enhance the values and assets you associate with floodplains. To give you a framework you can use to make decisions with that vast amount of information, we present six questions. The questions are short, but the answers must be comprehensive. Using this process could save time, money, and other resources once you decide on a course of action.

1. What Values or Assets Do You Want to Protect or Enhance?

It has been said that asking the right question is half the battle to getting a useful answer. Framing this first question is critical or the others have little meaning, because this is what you will use to evaluate any ensuing project. The most useful answer will be a clear, concise phrase that has a measurable component. It is more important to have the correct question with a vague answer than to have a precise answer to the wrong question.

For a homeowner, the answer might be, "To protect my home from flooding while preserving my view of the river." It's a short and measurable statement. Often, though, we answer this question with a predetermination of what activity we want to take: "To build a levee so my home doesn't flood." A levee might not be the best choice to keep your home from flooding, but by answering the question that way, you have already limited your options. What is important at this point is to not narrow the choices you consider. Keep the result, not the activity, in mind.

Though we often think only of problems in floodplains, there are also ways to enhance the amount or quality of water in our rivers. The six questions can also help you decide among various projects for enhancing water resources.

2. What Are the Apparent Risks or Opportunities for Enhancement?

In the world of natural resource management, hazard and risk mean different things. In short, the hazard is what can go wrong and the risk is how much damage might be done. For our homeowner, the hazard is flooding; the risk is the value of what she would lose if her home were destroyed.

This is an important concept because it lets us compare widely divergent assets and values. For a city, the hazard is flooding and the risk might be the value of downtown businesses that would be shut down during a flood. Or the risk might be the value of the water treatment plant that would be covered in a flood. Or the risk might be the value of tourist business lost because of the flood.

In some ways, risk allows us to discuss scale. It might seem that a homeowner would always lose in a discussion of scale, because one home's risk will be less than the risk to a city or the risk to local agriculture. However, there is rarely just one house at risk and damage to a conglomeration of houses (in other words, a neighborhood) can be just as much of a risk to a city as damage to commercial buildings.

Scale also brings us another dimension, and that is the importance of a given asset. To a homeowner, the complete protection of her home is crucial because it is probably her largest investment and carries emotional ties that few other things have. On the other hand, a city might specifically locate parks in a floodplain with the expectation that the equipment will be periodically destroyed but can be relatively easily replaced.

3. What Is the Range of Risk-Reduction or Opportunity-Enhancement Strategies Available?

Too often, we put on blinders and go back to what we know rather than evaluating what is available. Dams, levees, and dikes are not the only flood management options. A neighborhood that puts up a levee to protect itself may create floods in new areas downstream. As the dynamics of the river change, the levees may need to be reinforced periodically; if the river breaches the levee, the floodwater can be trapped in the neighborhood, unable to get back to the channel. This prolongs the time homeowners cannot access their homes.

For our homeowner's neighborhood, other options may include an upstream retention pond, changes throughout the basin to allow more water to infiltrate the earth (gravel parking lots instead of paved, for example, or elevating homes above the floodplain), or removing some houses to let the river recover some of its natural floodplain. In the end, some of these options will be cheaper than the construction and maintenance of a capital project like a levee because they take advantage of working with the natural processes of the river.

4. How Well Does Each Strategy Reduce the Risk or Enhance the Resource?

Our understanding of the environment is constantly improving. Many years ago, we spent millions of dollars to remove trees and boulders in river channels to produce "better" habitat for salmon and to reduce flooding risks. Money is now being spent to replace trees and obstructions in stream channels because we now know that we were wrong: the things we tried to fix actually worsened with our efforts.

For our homeowner, the effort of figuring out exactly which strategy works best is often beyond the time and resources she has. Usually, the homeowner can only address the consequences of flooding and only as her property is affected. That's why floodplain management is often best as an instrument of public policy. Usually, it's a government responsibility to coordinate an up-to-date list of policy options and provide a mechanism for deciding which competing interests should be addressed.

5. What Other Risks or Benefits Does Each Strategy Introduce?

For our homeowner, one drawback to using the levee to reduce flooding is that it will impede her view. Even if her house is high enough to see over the levee, such a structure disrupts the riparian zone along the banks of the stream. Levees are generally covered with grass so they can be more easily inspected, substantially reducing the diversity of flora and fauna, leaving fewer opportunities for watching wildlife, catching fish, or enjoying the procession of seasons as the leaves change on a variety of trees. Levees can result in greater velocities in the river, leading to greater erosion, which in turn requires increased levees. This is one reason that long-term maintenance costs must be included when estimating the total cost of building a levee. In the end, these costs often exceed the value of the protected land.

Another drawback is that the levee may increase the risk from larger but less frequent floods. When a levee fails, the force of the river is then concentrated in that area, causing more damage on the flooded land than if the levee did not exist. Levees can also impede water from returning to the river channel after a large flood. In 1927, the Mississippi River flooded catastrophically, leading to levees being overtopped and breached. Some areas had water above flood stage for two months.

It is an unfortunate truth that not everyone can win in every situation. In some cases, the homeowner may live in such a hazardous area, with such frequent flooding, that she must move. The city can buy her out, she can elevate her home, she can keep collecting on flood insurance claims until FEMA changes its policy, or she may just get tired of the damage to her belongings and decide to move. When there are winners and losers, our philosophy is to have the winners compensate the losers. For urban area floods, one strategy would be to store water in the higher watershed in order to manage the discharge. The winners—urbanites—could tax themselves, use flood easements to construct detention facilities, or purchase harvesting rights from upstream stakeholders.

As a society, we have often allowed those who build levees and other large structural projects to pass some of the costs on to other people, organizations, and even other generations. The long-term maintenance cost of levees is not always considered, even though they substantially increase the actual cost of a project. Dams can significantly change the behavior of water in the stream by creating a pond or lake behind them and reducing flow immediately downstream. Often dams are used to reduce flooding but in the process they destroy plants and animals that depend on flooding for nutrition or reproduction. They also restrict the amount of sediment that can move downstream, which can decrease the fertility of land and the amount of nutrients in the stream channel, and can cause downstream erosion. Scenic and recreational opportunities may be reduced along some stretches, even if they are enhanced in others. Fish and other wildlife may suffer precipitous declines if they can no longer get to their feeding, watering, or spawning grounds. Generally, we consider all these changes to be externalities and don't require builders to consider them or compensate communities for them. The true cost of capital construction projects includes all the changes the projects make. Many of those costs, however, are silently passed on to taxpayers or individual property owners.

6. Are the Costs Imposed by Each Strategy Too High?

Have you thrown out the baby with the bathwater? Sometimes it is just better to live with the risk. You move to a nice riverside community with beautiful views, nature walks, and fishing. It floods. The community suggests a levee district and a heavy tax to maintain the structure. The levee would block views and degrade natural processes, including the fishing hole, and may transfer erosive energy and deposit sediment downstream. The long-term solution may be worse than the original problem. A better strategy might be to elevate homes as opportunities became available, purchase flood easements upstream, and live with some flooding.

In some cases, a levee is still the appropriate answer. In other cases, a more regional approach is a lower-cost alternative. In many watersheds, the upper reaches of the stream have less development along them, being mostly farmland or forestland. A series of ponds in the upper watershed might keep our homeowner's house from flooding and would cost less in the long run. It would also keep more diversity of wildlife and flora along the riverbank. Finally, it would benefit a number of stakeholders along the river, including the ones downstream of our homeowner.

7. Make a Decision

After researching the problem and a range of solutions, and delineating the costs and benefits of each, it is time to make a decision. As you go through the process, the need for holding off on the decision will become clear. Solutions that seem easy often have unintended consequences. There are limited financial and political resources in the world. Floodplain decisions are usually based on a combination of the two.

When people use rivers and floodplains, they disturb the natural processes that produced the landscape. When making decisions about floodplain management, keep in mind the interlacing needs and results of physical, chemical, biological, and societal processes at work. Each will react to a change in the river, and each change will affect the others. Use the six questions in this book to figure out the lowest-cost, most beneficial, and longest-lasting solutions.

Case Study: Louisa County, Iowa

The Louisa 8 levee along the Iowa River, just above its confluence with the Mississippi, was built in the 1920s to protect agricultural land. Between the time of its completion and the 1993 floods, the levee breached every four years on average. Though floodwaters in an undeveloped river generally fertilize the land they cover, levee failures are different. Levees are constructed mostly of sand covered by clay, not the natural silts and clays that rivers normally deposit on their floodplains. The 1993 levee breach of Louisa 8 left scour holes and sandbars (sometimes more than 3 feet high) throughout this premier agricultural soil. This effectively prevented agricultural use unless the sand was removed or tilled into the naturally productive alluvial soils. The estimated cost of cleaning up the fields to return them to even a minimal level of agricultural production totaled nearly $3 million. The repair of this levee, even with a government subsidy, would have cost the levee district over $1 million.

Many assumed that the levee would be repaired, probably with an increase in height and strength. However, beginning with this flood, the federal government began to develop a process to review alternatives to business-as-usual levee repair. Some saw an opportunity to remove the levee and provide an outlet for the river at flood stage. In addition to a different form of flood control, this land could then be used as wildlife habitat. About 40 percent of the North American birds that migrate use the Mississippi Flyway.

1. What Values or Assets Do You Want to Protect or Enhance?

The asset here was the land itself. Flood damage severely degraded the ability of the land to grow crops. The levee's history of breaching did not bode well for expected increases in flooding due to watershed and/or climate changes.

Other riverine values that could be enhanced were increases in scenic values, multiple recreation uses, and wildlife habitat. This could be accomplished by changing flood management practices and using the land for temporary storage at flood stage, allowing the land to revert to a more natural state.

2. What Are the Apparent Risks or Opportunities for Enhancement?

The land could not support commercial agriculture without the levee. However, even before the 1993 floods, agricultural benefits were marginal. The levees destroyed any semblance of the natural state of the river. Predicted increases in severe storms because of climate change would stress the levee even more.

Societal values have changed since levees were first built in the 1920s. At that time, the consensus held that levees should be built to keep the river from flooding. Since then we have come to better understand the risks associated with levees, from possible failure to destruction of wildlife habitat. Levees are no longer seen by all as the answer to flooding.

3. What Is the Range of Risk-Reduction or Opportunity-Enhancement Strategies Available?

The USACE intended to rebuild the levee.

Another option became available after the 1993 floods. Because of the widespread catastrophic flooding, the U.S. Department of Agriculture (USDA) offered a program to enhance wetlands by taking land out of agricultural production. Called the Emergency Wetland Reserve Program, it offered to pay landowners for a permanent easement on their property. The easement prohibits commercial agriculture, but allows planting food for wildlife.

4. How Well Does Each Strategy Reduce the Risk or Enhance the Resource?

Past experience showed that the levee did not adequately protect the land for farming. Returning the land to wetlands removes the land from agriculture, but enhances other potential uses of the river.

5. What Other Risks or Benefits Does Each Strategy Introduce?

The USACE initially perceived an unrepaired levee element as a break in an otherwise controlled system. The processes of building and maintaining levees are well understood.

Allowing a stretch of river to revert to wetland has some long-term maintenance costs. It also removes productive land from county tax rolls.

Because levees move water downstream faster, they can increase downstream flooding. By removing levees, the increase in flood-stage detention could reduce downstream flooding in addition to restoring natural resource values.

6. Are the Costs Imposed by Each Strategy Too High?

If the land was needed for agricultural production, a levee was necessary. If the land is more productive (in a societal sense) as a wetland, then rebuilding the levee removes that option. A choice is needed to decide the future use of the land.

7. The Decision

Changing the mind-set of business as usual took many parties working together. The landowners worked with FEMA, the USDA, and the Iowa Natural Heritage Foundation staff on possible options. USDA and FEMA developed innovative funding mechanisms to supply most of the project funds; the Iowa Natural Heritage Foundation made available funds derived from the federal Duck Stamp hunting program.

Levee district taxes had increased each time the levee breached, levee repairs were made, and the devastated farmland was restored. These costs were making farming less and less economically feasible.

The farmers who managed the Louisa 8 Levee District carefully considered their options at several meetings and eventually determined that the levee district would not accept USACE funding to repair the levee. Instead, the affected landowners sold easements to the USDA Emergency Wetland Reserve Program and the levee district dissolved. The original planned use of the land was a privately operated hunting preserve. Later, many of the farmers voluntarily went a step further and sold their land for use as a public open space. The cost of the project was about $2 million, far less than it would have taken to recover agricultural land from the debris of the levee and repair the levee.

Past owners lost tillable land and the county lost some tax base. However, factoring in the cost of maintaining the system and the increase in natural resource values, the benefits are proving to be much greater than the costs.

Now called Horseshoe Bend, the former Louisa 8 levee district is part of the Port Louisa National Wildlife Refuge. The area includes grassland, wet meadow, seasonal, and semipermanent flood-emergent wetland habitat. Bottomland trees have been planted. The variable habitat supports a wide range of wildlife species. The wetland complex provides flood storage, fish passage, and spawning. Shorebirds, waterfowl, wading birds, and grassland bird species use the land in migrations. Several bald eagle nesting attempts have occurred. Two breaches in the levee allow the river to reconnect to the floodplain. In high water periods, the water can flow into and out of the complex at will.

This project offers a model for dismantling additional individual levees and levee systems. As societal values change, we increasingly treasure a more complex riverside system. We now see recreation, flyways, and other wildlife habitat as important systems to maintain.

As our climate continues to change, forecasters expect more extreme storms and longer summers. The twin problems of water scarcity and increased flood heights will require increased opportunities for water storage that can recharge aquifers and detain floodwaters.

Chapter 2

A New Vocabulary

Floods are more than just hazards. It is more constructive to think of floods as a natural process of rivers that provides both risk and opportunity. This allows us to think of managing floodwater for our benefit. To maximize these benefits, we need to understand resiliency and maintain it in our riverine systems. In addition to understanding physical, chemical, and biological systems, we must also be able to assess our societal capabilities. Three important federal actors in flood control are the U.S. Army Corps of Engineers, the Natural Resources Conservation Service, and the National Flood Insurance Program. The National Marine Fisheries Service, through a recent biological opinion, may be a new participant in floodplain management programs. The No Adverse Impact initiative, from the Association of State Floodplain Managers, calls for new approaches to flood management.

A New Vocabulary

Floods are hazards: That's the traditional view. It's true they can be disasters as we saw along the Mississippi River in Chapter 1. However, flooding can also be beneficial by replenishing aquifers and agricultural fertility. We need a greater cultural acceptance of the beneficial effects of floods. They provide opportunities to store water; reduce channel migration, erosion, and downstream flooding; and recharge aquifers. These positive effects will become increasingly important in some areas as our climate changes. To help understand that floods bring opportunities as well as risk, we'll need a new and more objective word than hazard.

Here is some of the vocabulary we'll use in this book, along with basic principles and key actors in the management of floods. We start with the definition of a flood and its potentially harmful effects, and then introduce terms reflecting the positive effects floods can have.

Flood

In itself, a flood is neither positive nor negative. It's simply water spilling over its banks. At its most basic, a flood is a change or disturbance in the river.

Floodplain

In this book, we're concerned with the floodplain that is physically, hydrologically, or biologically connected to the main river channel. There are many definitions of a floodplain, but they all include land that the river periodically covers.

The 100-year flood is the basis of many plans. On average, these floods happen every 100 years, but they can happen in consecutive years or 200 years apart. There is a 1 percent chance of this flood happening in any given year. A 500-year flood is a similar concept, but has a 0.2 percent chance of happening in a given year. There are many problems with this terminology, as will be described in the next chapter. One of the most significant is that these terms assume that landscape characteristics do not change.

Hazard

A hazard is something that can go wrong. In this respect, floods are hazards; they can definitely cause damage.

A hazard has a potentially adverse effect.

Risk

Risk is how much damage a hazard might do. For a homeowner, the hazard is flooding; the risk is the value of his home if destroyed. When we talk about floods, we generally think of large events that damage something and ignore the small events that are neutral or beneficial. It's easy to understand why. Most of us live in an urban environment where floodwaters are nearly always a problem.

Let's consider risk a little further. Most professions seem to define risk in a similar way: Risk is a relationship between consequences from a hazard, its frequency, and the capabilities available to either reduce the effects or lessen the threat. There is one profession that sees risk in a different way: psychologists. They address risk as perception. The perception of risk has to do with control, level of scariness, and maybe some

DNA hardwiring. Many people feel more comfortable (less at risk) when they are the one driving a car and can control speed, steering, and so on. The 9/11 attacks were more frightening than tobacco, even though tobacco-related lung cancer causes many more deaths per year than the attack on the World Trade Center. It is important to separate the perception of flood risk from what actually happens.

Risk is the amount of potential damage from the change (in this case, a flood).

Capabilities

Assessing a level of risk also involves capabilities. If a river suddenly changes course and affects your home by flowing inside, the increased risk would partly depend on how capable you were to react to the change. Your risk would be extreme if you were in a wheelchair or asleep; it would be lower if you were able to outrun the rushing waters.

A capability is the amount of resources available either to reduce the change (the flood) or the effect (the risk), or to enhance water resources (the benefit).

Disaster

Flooding and drought are inevitable. Disasters aren't. This is our main point: Floods need not be disasters and we can even benefit from floodwaters. Potential damage can be minimized in many ways.

If the risk is realized (that is, if there is a flood that destroys property or causes deaths), it is a disaster.

Benefit

Benefits are positive effects from changes—in this case, the good things that come from floods.

Opportunities

If you're a farmer or rancher, you may depend on that change (the flood) to fertilize your fields or provide water for herds. If you depend on an aquifer for drinking water, floodwaters may recharge it so water is continually available.

Our (rightful) aversion to flood damage makes us perceive floods as hazards rather than simple natural processes. However, if we look at the total effect of flooding we come to a different perception. Flooding is a necessary part of the water cycle we rely on to survive. We may want to change the placement or timing of floodwaters, but we need the water to be in the system.

By using the time between floods to analyze the watershed and develop our priorities, a flood can present an opportunity to implement long-range plans that have been developed by a community. Without such planning, recovery from floods—especially major flood events—can be chaotic and benefit only the opportunistic, not necessarily the community at large.

An opportunity is the opposite of risk. If opportunities are realized, they are a benefit.

Same Event, Different Outcome

Using this vocabulary, we can dissect the effects of a flood.

1. *Change*: A stream periodically changes by flowing over its banks.
2. *Risk*: A house may wash off its foundation.
3. *Capability*: Theoretically, the house may be elevated above the floodplain, but the homeowners can't afford it.
4. *Disaster*: The house is not elevated and is destroyed.

Now take the same scenario and add a capability. In this watershed, wetlands and marshes may be restored to provide water storage. The same event then becomes:

1. *Change*: A stream periodically changes by flowing over its banks.
2. *Risk*: A house may wash off its foundation.
3. *Capability*: Improved wetlands give water a place to collect, slowing the flow of water through the river channel and lowering the height of floodwaters.
4. *Benefit*: Urban flooding is reduced, homes are saved, and the wetlands also serve as parks and wildlife havens.

For these reasons, we'll define risk as the combination of a change, the potential harmful effects, and your capabilities. An opportunity is the ability to use your capabilities to react to the change and get a benefit. A disaster is a realized risk and a benefit is a realized opportunity.

This does not mean that every flood can be a benefit everywhere. Flooding in an urban area is frequently disastrous, but it does not have to be so. We can neutralize the danger in some places, and perhaps focus the benefits in others.

Resiliency

The ultimate goal is to have a resilient system—that is, a floodplain, river, and watershed that can tolerate changes like floods. A resilient system will recover to its approximate original state once the disturbance is over or removed. Figure 2.1 shows that resiliency is somewhat dependent on the size and complexity of a natural system. A watershed is naturally resilient because it is such a large, complex web of flora and fauna. On the other hand, the habitat of a single animal species might be fragile if the animal eats only certain kinds of plants or can live only in a narrow range of temperatures.

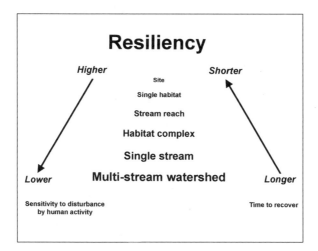

Figure 2.1
Different scales have different levels of resiliency. A new development might destroy the habitat of a specific plant or animal, but have little effect on the watershed as a whole. Conversely, it might be relatively easy to restore the specific habitat, but once the watershed seriously degrades, the recovery time (and needed resources) will be long term. Effects accumulate through the watershed so that a number of small projects can degrade the watershed as much as a single enormous development.

Table 2.1

Ecological Resilience Is a Better Framework Than Engineering Resilience

Engineering Resilience	*Ecological Resilience*
Seeks stability	Accepts inevitability of change
Resists disturbance	Absorbs and recovers from disturbance
One equilibrium point	Multiple, nonstable equilibria
Single acceptable outcome	Multiple acceptable outcomes
Predictability	Unpredictability
Fail-safe	Safe-fail
Narrow tolerances	Wide tolerances
Rigid boundaries and edges	Flexible boundaries and edges
Efficiency of function	Persistence of function
Redundancy of structure	Redundancy of function

As a society, we need to be able to withstand periodic large floods and rebuild as necessary without causing massive societal, political, or ecosystem upheavals.

Resiliency, however, can mean many things. Typically, flood management projects have used an engineering definition of resiliency: The rate at which a system returns to a single steady state or cyclic state following a disturbance. We suggest using an ecological definition: The amount of disturbance a system can withstand before it changes to a new set of reinforcing processes and structures.

This is a dramatic change in outlook, illustrated in Table 2.1. *Engineering resiliency* is looking for a stable system that has little variation. It is often predicated on a built environment with a fail-safe outlook—that is, that there is enough redundancy so that the system will not fail. (Of course, if it does fail there is often a greater catastrophe than if the engineering were not in place.)

Ecological resiliency starts with a different premise: that natural systems are always in a state of flux. They should be able to function within a wide range of conditions. Systems are designed to be safe-fail, meaning if they fail, it will not be a catastrophe.

Risks and Opportunities

Both risks and opportunities are the result of the combination of the original change, potential effects, and our capabilities, as illustrated in

Table 2.2
Floods Present Complex Risks and Opportunities

	Risks and Opportunities (Result of change, effect, and capability)	
Focus on change *Primary and secondary*	*Manipulate the effect* *Beneficial or adverse*	*Capabilities*
Hazard or disturbance	Vulnerability or exposure	Strategies made up of approaches and tools
Location Frequency Severity Timing	Systems: Built environment Natural environment	Money Power Timing

Table 2.2. Risk reduction (or opportunity for enhancement) has two sets of capabilities. We can:

- Focus on the change.
- Manipulate the effects.

An example of a potential change might be a rotten tree standing in a forest. This tree may be huge, located on a steep hill, and ready to fall down any second. The tree is a hazard but not a risk. If we step under the tree, there is now a risk: something of value (us) can be damaged. This risk is also influenced by our individual capabilities. If you are deaf and can't hear the tree crack, don't have health insurance, or are far away from a hospital, your risk is higher.

In addition, this rotten tree is an apple tree but the apples are high off the ground. These apples represent an opportunity. If we don't have the capability to retrieve the apples, there is no benefit. However, if the tree falls and the apples are now on the ground, we realize a benefit.

Primary and Secondary Changes

It is helpful to distinguish between primary changes and secondary changes (those caused by primary changes). Using our rotten tree example, there may be a power line near the rotten tree and when the tree falls, the power line may cause a fire. In this example, the tree causes a change and the thing of value would be the power line. The falling power line would be a secondary change. It is a hazard in its own right and might affect other things we value. It might also be an opportunity, if the grasses below it would benefit from fire.

It is helpful to think of such a sequence of events as primary and secondary changes. This distinction will help us focus on the root causes of damage and this in turn will help isolate risk-reduction and opportunity-enhancement measures.

For example, a flood causes a change and a risk to our object of value, a manufactured house in the floodplain. If the flood forces a house off its foundation, the floating house or its debris may become a secondary hazard to neighboring houses. By focusing on the root cause of the problem (the floodwaters) and what was initially affected (the original house) we can better direct our risk-reduction efforts. In this case, the risk caused by floating debris can be reduced by anchoring the house to the land as is required of manufactured homes by the National Flood Insurance Program (NFIP).

Secondary hazards to flooding can be landslides, which are a risk as soil becomes saturated and less stable. Flooding can also cause fires, because flooding can extinguish home furnace pilot lights, allowing gas to collect in voids and then spark.

There is a complicated relationship between flooding and wildland forest fires. Forest fires can exacerbate flooding when they bake soils, causing them to become less permeable, which makes water run off more rapidly. Soils with less moisture mean trees with less moisture, which increases the danger of fires, which can in turn increase the number or severity of floods.

As will be discussed in subsequent chapters, climate change may alter the flow characteristics of many streams. Some models suggest that many places will have reduced snow but heavier rain in winter due to climate change. Streams now heavily influenced by snowmelt may have reduced summer flow. Figure 2.2 illustrates how one type of stream might react to future climate change. Snow-dominated streams tend to have peak flows in late spring and summer when the snow melts. As temperatures warm, less snow will fall to form snowpacks. These streams may move into a mixed regime (rain and snow). This produces peaks in both winter (rain) and spring (snowmelt) months. If temperatures continue to warm, the result will be little or no snowpack and streams will be fed primarily by rain. This produces high flows in the winter, but very little flow in summer months when water is needed for drinking and irrigation. Not every stream will follow this pattern of change—climate change will affect different areas in different ways, many of which are not yet apparent.

Risk can be reduced by focusing on either the change or the effects. We have often tried to reduce flood risks by focusing only on the change (the flood). We have attempted to control the river by building massive structures such as dams and levees. We have armored riverbanks and

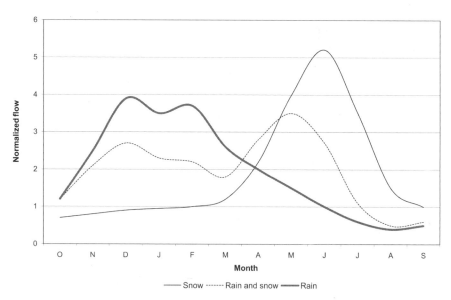

Figure 2.2

Three hydrographs show how streamflow might change with climate change. In its original state, the stream is dominated by snowmelt. Warmer temperatures may reduce the amount of snow, leading to a rain and snow dominated regime. Still higher temperatures might eliminate snowpack, leaving a rain dominated system with very different flow and flood patterns.

straightened river channels. These structural measures sometimes caused secondary disturbances that were greater than the primary ones. Big projects with big footprints may be beneficial over the short term but represent a much-increased long-term risk.

More cost-effective approaches often have a smaller footprint. They reduce the risk by focusing on what is harmed, not by controlling the flood. As an example, compare the long-term costs of elevating a house above the floodwaters to building, maintaining, and accommodating the indirect costs of a dam attempting to control the river.

Hazard as Change

Because flooding drives both risks and opportunities, causing both disasters and windfalls, it is too simplistic to call flooding a hazard. We need a better word—a more neutral word. Biologists use the word disturbance instead of hazard to reflect a temporary change that may be positive or negative. However, many also view disturbance as negative so we suggest the word *change*. We'll reserve the word *hazard* for a change that causes harmful effects.

When we profile change, we look at location, frequency, severity, and timing.

- Location: geographic location, spatial distribution.
- Frequency: probability, seasonality, temporal distribution.
- Severity: magnitude, duration, spatial extent, physical mechanism.
- Timing: speed of onset, warning time, preparation time.

Location is an important characteristic of change. Rivers have a place—a location. It may be fairly stable or may be more volatile in areas where the geology is evolving, such as Alaska. Just as the river has a location, so does its watershed and floodplain.

Probability is an important measure of *frequency*. If a flood has a return interval of 200 years, it might not be cost-effective to put money into solving the root cause. If the flood occurs every couple of years, or is steadily increasing because of increased development in the watershed, then it *is* cost-effective to spend money on the problem.

When we think of the *severity* of change in terms of flooding, we usually think of the depth of the water on the ground. However, it can also include changes in water velocity or the debris transported by floodwaters.

Timing is an important element of risk and opportunity. During the floods on the Mississippi described in Chapter 1, many communities had weeks of warning as the high flow worked its way to the Gulf of Mexico. In other circumstances, communities located along shorter streams in mountainous areas may have only minutes to respond.

Beneficial and Adverse Effects of Change

Change creates effects, vulnerabilities, and exposures. Dennis Mileti, in *Disasters by Design*, lists three vulnerable environments. Each has its own systems, mechanisms, feedbacks, and resources:

- *Human-built*: including buildings and infrastructure.
- *Natural*: including salmon runs, soil, and trees.
- *Societal/political/organizational*: including institutional capacity, population distribution, and transportation systems.

The effects can be injuries to people or damage to property and infrastructure. There are economic consequences—lost jobs, business earnings, and tax revenues, as well as indirect losses caused by interruption of business and production.

It is helpful to understand the relationship of exposure to vulnerability. It is similar to the relationship between hazard and risk. A flood may cover two subdivisions equally. Both subdivisions are equally exposed to the floodwaters. One subdivision, however, may include only homes that have their first floors elevated above the flood level. The other subdivision may include homes that were built at ground level and below the base flood elevation. Both subdivisions are equally exposed, but the elevated one is less vulnerable to the floodwaters than the subdivision in which homes were not elevated.

Though we often think of vulnerability as being hazard- or change-specific, there are populations that are inherently vulnerable to almost any change. These societal, political, and organizational vulnerabilities are as important as those in the human-built and natural environments, but are often not included in vulnerability assessments. In many parts of the United States, floodplain sites have been among the most affordable locations available to the very poor. Efforts to purchase these flood-prone properties from their disadvantaged owners have been made with the best interests of the floodplains and their residents in mind, only to meet resistance because alternative low-cost housing is unavailable for many.

Philosophies of Flood Management

Three federal agencies historically concerned with flood management are the U.S. Army Corps of Engineers (USACE), the Natural Resources Conservation Service (NRCS), and the Federal Emergency Management Agency (FEMA) through its administration of the National Flood Insurance Program (NFIP). They are not the only agencies that deal with flood control, but each of them highlights a different approach to managing flood-related change. Each spends enormous amounts of money each year on flood management issues, so it is important to understand their philosophies.

The Department of Commerce (through the National Marine Fisheries Service) and the Department of Interior (through implementation of the Endangered Species Act) may have important future contributions to floodplain management. The addition of federal agencies concerned with biology highlights the complex nature of floodplains and our evolving understanding of them.

The Association of State Floodplain Managers developed the No Adverse Impact (NAI) initiative, a vision of what river management could look like along with suggested capabilities to realize it.

U.S. Army Corps of Engineers: Managing Change, Often with Big Footprints

Protecting communities from flood dangers often falls to the U.S. Army Corps of Engineers. The USACE is a principal federal agency responsible for large structural works such as straightening channels or building dams and levees. These projects reduce flooding through manipulating the location, frequency, severity, or discharge timing of the river. The USACE's tools are often narrowly defined to be within the context in which they have regulatory responsibility—manipulating changes in the river channel. Addressing future development, land cover, upland storage, and climate change are not USACE capabilities.

The USACE has based construction on an economic and environmental risk analysis in which significant past events represent the key factors. By law, the USACE must consider the most probable flood (often based on the largest flood of record) and determine whether the benefits outweigh the costs of construction. Future development conditions of the watershed and the potential effects of climate change are not factored into their analyses.

It is important to realize that, like other federal agencies, the USACE is politically responsive (providing a potential risk and benefit). A colonel usually heads each USACE district office. To be promoted to general, an officer is nominated by the president of the United States on the advice of the secretary of defense and confirmed by the Senate. This can place a lot of political pressure (a risk) on well-serving and dedicated USACE employees. It can also be a benefit because senior USACE officers must have an understanding of the political process.

The USACE also supports flood management under the Flood Control and Coastal Emergency Act. The USACE can undertake activities including disaster preparedness, flood response, and rehabilitating flood control works threatened or destroyed by flood. This includes the Rehabilitation and Inspection Program (RIP), which is directed largely at assuring levee reliability. If a levee system is in RIP before a flood and owned by the federal government, USACE will restore a damaged levee to its preflood state at no cost. It will charge a nonfederal owner only 20 percent of the total cost.

Participation in RIP requires periodic USACE inspection of the levees. Inspection policies create an enormous problem for river systems. USACE regulations generally require removing trees that are more than 2 inches in diameter, though this can be amended by individual districts. For example, the Seattle district allows trees up to 4 inches. In either case, the absence of medium- and large-size trees severely affects the entire riverine biota. As a result, most levees are grass-covered and the kind of complex natural web of plants and animals that provides some natural flood protection is actively discouraged. These policies continue, even as scientific research demonstrates that tree roots and brushy vegetation often decrease the likelihood of levee failures.

The USACE also administers the wetland permitting program under Section 404 of the Clean Water Act. Removal or reduction of wetlands can significantly change a river system.

Funding for large USACE projects is typically included within specific legislation. It is important to remember that as part of the army, the USACE also plays a role in our nation's defense, and its resources are stretched when we are at war.

Natural Resources Conservation Service: Smaller Footprints, but Cumulative Effects

Similar to the USACE, the Natural Resources Conservation Service (NRCS) focuses on managing the change (the flood) to reduce risks and take advantage of opportunities. Unlike the USACE, the projects of NRCS are typically small, but their influence should not be underestimated. NRCS offers an example of how numerous small projects, done individually, can have enormous positive or negative cumulative effects in changing a watershed. NRCS provides technical and financial assistance to America's private landowners and managers. Participation in its programs is voluntary. Its conservation programs help people reduce soil erosion, enhance water supplies, improve water quality, increase wildlife habitat, and reduce damage caused by floods and other natural disasters. Its focus on conservation parallels this book's focus on working with natural systems and processes. However, because NRCS projects *are* generally small, their cumulative effects on a watershed may not be taken into account.

National Flood Insurance Program: Focusing on the Effects of Change

The National Flood Insurance Program (NFIP) is based on the principle that making insurance available will provide an incentive to reduce flood damage. However simple in concept, application of the program has become very complex. More complete information about the NFIP, including a case study, is in Appendix A.

The NFIP has three major components.

Insurance

Flood insurance is available in communities that require new and substantially damaged buildings to comply with flood risk-reduction measures. As long as a community is participating in the NFIP, flood insurance is available both within and outside of floodplains.

Flood insurance is required if the federal government is involved in the financing or building of a structure, and the land is located within a designated 100-year floodplain. Subsidized flood insurance provides the incentive to comply with building regulations. Property owners purchase coverage for contents and structures from private insurance companies and the federal government underwrites the risk. Policy premiums support administration of the program.

Regulation

Local governments must adopt and enforce ordinances to restrict development within designated flood hazard areas. These regulations should eliminate or minimize future flood damage for new and substantially improved structures within a designated 100-year floodplain.

Floodways are important to the NFIP. They are regulatory creations, not defined by scientific parameters. Floodways are linear areas set aside to convey the 100-year discharge without increasing the base flood elevation by more than 1 foot. Floodways usually include the river channel and some overbank areas. The floodway designation assumes a built-out condition across the floodplain that results in a 1-foot rise in 100-year flood elevation. The encroachment on the river corridor that produces this rise defines the cross section of the floodway.

Mapping

FEMA prepares floodplain maps that reflect existing conditions, not considering projected development. New and substantially improved structures located within this mapped area are regulated. A Special Flood Hazard Area (SFHA or Special Area) is an area within a 100-year floodplain delineated on floodplain maps.

There are two Special Area maps. (An example is in Appendix A.) One is very general and often just a best guess at illustrating the 100-year floodplain. These Flood Hazard Boundary Maps (FHBMs or hazard maps) are a best stab at describing the Special Areas. They were first prepared in the 1960s and 1970s and have largely been replaced by more detailed maps. However, they may still be the only maps available in rural areas with little expected development.

Flood Insurance Rate Maps (FIRMs) have largely replaced these hazard maps. These maps are the product of rigorous analysis and contain more detailed information. FIRMs illustrate the 100-year floodplain boundary—the actual elevation that floodwaters are expected to reach during a 100-year flood—and often include the area expected to be inundated during a 500-year event. In urban or urbanizing areas, FIRMs also illustrate the floodway.

Potential Problems

Several problems arise from NFIP regulations and practice. These include:

- Damage not defined as substantial does not trigger regulation. Floods can damage 49 percent of a building year after year, and the NFIP will continue to pay claims without requiring mitigation measures.
- Extremely risk-prone properties may receive payments on claims totaling many times the value of the property.
- NFIP regulation is community specific. Flooding, however, is not community based but is a response to cumulative actions within the entire watershed.
- Downstream effects aren't addressed by the NFIP. It encourages the use of fill to elevate structures. Through its dependence on floodways, it limits opportunities to spread water over the land. It also misses opportunities to increase flood storage, dampen energy, and expand lateral biological continuity.
- Hazards and opportunities outside mapped areas are not addressed.
- Beneficial natural processes and values are not addressed.
- Not all flood-related disturbances are caused directly by the rise and fall of floodwaters. Other problems include the effects of velocities, long-standing water, channel migration, and effects from sediment.
- Special Area maps are intended to reflect current conditions but are

often outdated before they are adopted. At best, they reflect conditions at the time the surveys were undertaken. They don't acknowledge future conditions, such as increased runoff and loss of storage due to basin development. This is of particular concern because communities use these maps to make long-range planning decisions, but they should not.

- Insurance premiums are based on mapped information, not true risk. In communities with rapid changes in land use and land cover, insurance rate maps become quickly outdated and no longer reflect realistic risks.

Community Rating System

An optional program offered by the NFIP is called the Community Rating System (CRS), which encourages and recognizes floodplain management activities that exceed minimum NFIP requirements. Flood insurance premium rates are discounted up to 45 percent to reflect actions meeting the three goals of the CRS:

- Reduce flood losses.
- Facilitate accurate insurance ratings.
- Promote awareness of flood insurance.

More than a thousand communities of all sizes receive flood insurance premium discounts. Discounts are one benefit of participating in CRS, but it is more important that these communities are carrying out activities that save lives and reduce property damage. They represent a significant portion of the U.S. flood risk because they include over 67 percent of the NFIP's policy base.

Endangered Species

Endangered species such as fish, amphibians, other animals, and plants are often affected by flood management practices. A 2008 National Marine Fisheries Service biological opinion clearly tied the flood management practices supported by the NFIP to decreased fish habitat and decreased ability of watersheds to regulate floodwaters.

This opinion was required by a 2004 federal court decision and related to the Endangered Species Act (ESA), which is administered by the U.S. Fish and Wildlife Service of the Department of Interior. This specific biological opinion addresses rivers draining into Puget Sound, but it may have a much broader impact. The suggestions offered in the opinion to support salmon and orcas are the same being offered in this book to reduce flood damage. Chapter 10 of the biological opinion gives specific recommendations by NMFS.

From the report:

> The importance of floodplain habitat to salmonids cannot be over-stated. In the Skagit and Stillaguamish Basins, more than half of the total salmonid habitat is contained within the floodplain and estuarine deltas, while this habitat encompasses only 10 percent of the total basin area (Beechie et al., 2001).
>
> Human influence has degraded watersheds and wetlands, diminished the amount of available floodplain, and degraded remaining intact floodplains throughout Puget Sound. Floodplain impacts include the direct loss of aquatic habitat from human activities (filling), disconnection of main channels from floodplains with dikes, levees, revetments, and roads, and reduction of lateral movement of flood flows with dikes, roads, levees, and revetments. For example, King County [Washington] is responsible for maintaining 119 miles of levees that line six river systems. In many stretches the levees are a mosaic of bushes and shrubs, including invasive plant species, that have low ecological value for salmon habitat. Removing vegetation from levees decreases riparian functions that enhance salmonid habitat. . . .
>
> Three elements of the NFIP (floodplain mapping, minimum floodplain management criteria, and the CRS) directly and indirectly lead to changes in floodplain environments and eventually to floodplain development. These changes in the floodplain environment adversely affect the habitat and habitat forming processes for listed species in the Puget Sound region. . . .
>
> By minimizing future habitat losses and by utilizing its authorities to encourage the restoration of floodplain habitat through the removal of structures and other measures where feasible, FEMA can both avoid the likelihood of jeopardizing listed species through NFIP implementation and fulfill the NFIP's purpose of reducing the risk of flood losses by encouraging land-use practices that constrict floodplain development.

No Adverse Impact (NAI)

An initiative can be defined as a structured strategy promoted by government and nongovernment alike to address some shortcoming. The No

Adverse Impact (NAI) initiative has provided a substantive base for community groups organized to preserve and enhance their riverine values.

In the 1970s, those involved with floodplain management in the Midwest created the Association of State Floodplain Managers to support and guide the emerging NFIP. The Association developed NAI in the late 1990s. This initiative addresses some of the shortcomings of the NFIP, including the lack of a component to protect the environment, lack of attention to new development, and lack of focus on projects in the upper watershed to reduce flood hazards. The NAI initiative focuses on supporting incentives that reduce adverse impacts *to* the development and adverse impacts caused *by* the development.

The following statement is from the Association of State Floodplain Managers NAI Tool Kit:

> Natural floodplains provide opportunities for open space, parks, recreation, habitat for wildlife and fish, hiking and biking trails, alternative agricultural crops, and add to quality of life. Flood levels do not increase over time in your community, because you use NAI approaches.
>
> Increases caused by any development are mitigated so they do not impact others. Development is done in a manner that does not pass the cost of flooding on to other properties, other communities or to future generations.

The NAI is driven by the recognition that individual actions on the floodplain can adversely impact others. Many of the recommendations in this book are based on that framework. The NAI should be expanded to include low-impact development initiatives and to incorporate measures that take advantage of natural processes to reduce flooding.

Case Study: Snoqualmie, Washington

Once a mill town, today Snoqualmie, Washington, is an exurb of Seattle. The town center is a blend of museums, a historic railroad yard, a wooden monument to ancient old-growth forests, and a dozen or so commercial structures.

Historic Snoqualmie has a population of about 1,700. Except for two homes, all of historic Snoqualmie—almost 700 units—is in a Special Flood Hazard Area. Mill workers themselves built these houses many years ago—simple cottages and bungalows constructed with sturdy old-growth timbers. A few postwar ranch houses replaced the more poorly built and flood-damaged homes. Over the past few years, land values skyrocketed and as a result, vacant lots were developed and older homes were upgraded. Downstream from town is Snoqualmie Falls, which creates a restriction that backs up water into town during floods. Snoqualmie was included in thirteen presidential flood-related declarations between 1965 and 2001, but the water is clean and slow-rising with little debris. A thousand or more acre-feet of water can be stored within the community for several days until it drains over the falls, reducing flood flows to downstream communities.

Snoqualmie is an NFIP community and requires new homes to be elevated above the base flood level, and existing homes to be elevated when substantially improved. NFIP allows communities to legislate and enforce stricter standards than it requires.

1. What Values or Assets Do You Want to Protect or Enhance?

The community wanted to enjoy small-town life along the river and protect their assets from flooding.

Downstream communities wanted lower flood levels.

2. What Are the Apparent Risks or Opportunities for Enhancement?

Life safety was not a major issue because of the slow rise of the water and its lack of sediment. The risks were the cost of frequent repairs, lost contents, and the nuisance of dealing with flooding. Another important asset was the money that could be saved through risk reduction. Frequent flood damage impedes economic development.

There is a possibility that flood frequency and stage might increase as the climate changes and flood storage is increasingly lost through development. Storing floodwaters within the city could become increasingly valuable to downstream users to reduce their flooding and provide water for agriculture.

3. What Is the Range of Risk-Reduction or Opportunity-Enhancement Strategies Available?

Traditional answers would include levees, a new dam, or widening the natural dam, allowing greater discharge over the falls.

Snoqualmie could relocate.

The city could encourage elevating buildings by adopting stricter standards than the minimum requirements of the NFIP and taking advantage of state and federal flood-reduction grant and loan programs.

4. How Well Does Each Strategy Reduce the Risk or Enhance the Resource?

Dams and levees are expensive to build and maintain and pose a potential catastrophic risk if they fail.

With 6,500 residents and few better locations nearby, relocation is not a realistic option.

The NFIP definition of substantial improvement could be amended to require elevating structures if less than the minimum standard of 50 percent of market value is improved. Structures could also be required to elevate above the minimum NFIP standard.

5. What Other Risks or Benefits Does Each Strategy Introduce?

Dams and levees have high long-term costs and can substantially change the characteristics of the watershed and river.

A benefit of requiring new and remodeled homes to elevate above the base flood level is to reduce damage to individual structures, as well as the amount of debris in floodwater to damage others. It would also reduce personnel and equipment costs of local emergency responders if they were needed less frequently for response, rescue, and recovery activities. Allowing water to pond in the area would reduce the downstream flood effects and would enhance benefits to downstream water users. This strategy would also help create a livable, attractive community that enjoys the benefits of the river but suffers minimal flood damage and doesn't create adverse impacts to downstream communities.

6. Are the Costs Imposed by Each Strategy Too High?

Even if levees could protect Snoqualmie, they would simply move the flooding downstream to another town. Levees would also remove many of the riverside values that residents enjoy.

Elevating homes does not end flooding and high water will continue to flow through town periodically, though the homes will no longer be at risk. There is still a cost to homeowners, however, because their homes may not be accessible during floods. This is a lower cost than if they were not able to get to their homes and also suffered damage to the homes and belongings.

7. The Decision

Snoqualmie decided to encourage homeowners to elevate their homes to a higher standard using federal and state funds whenever possible.

It costs less to elevate new homes than to retrofit them. At the time of construction, elevating a home 5 feet costs less than 5 percent of the total cost. Retrofitting existing homes may cost up to $50,000.

The first significant elevation/retrofitting project occurred in 1987 after a Federal Disaster Declaration. The city redefined substantial improvement to mean repair, reconstruction, or improvement of a structure that cost 10 percent or more of the building's market value. This definition far exceeds the minimum 50 percent set by the NFIP. (This threshold has since been raised above 10 percent, but the high standard provided great initial incentive.) About sixty homes were elevated between 1987 and the spring of 2002 and more than 100 new, elevated homes have been built.

After the 1987 Federal Disaster Declaration, the Small Business Administration made loans available to homes that met existing codes and about a dozen more homes were elevated. Another fifty homes were elevated through the Hazard Mitigation Grant Program (HMGP) after presidential declarations in the November 1995, February 1996, and Winter 1997 floods.

The HMGP program contributed 75 percent to 87.5 percent of the cost of elevating a home. The homeowner was responsible for the balance. However, for many homes the state of Washington covered 50 percent of the remaining costs after the grant. The city also received Flood Mitigation Assistance funding to elevate three homes with repetitive losses, requiring the homeowner to contribute 25 percent of the costs of elevation. As a result, few homeowners paid the entire cost of elevating their home.

In this way, Snoqualmie is able to preserve its historical heritage, promote civic and economic values, and continue to be a place where families want to live. It will always flood, but with every year the community is reducing the damage from flooding while enhancing riverside benefits.

Chapter 3
Rivers and Floodplains

Floods can swamp roadways, engulf buildings, and even trap people who then need rescuing by helicopters. However, floods also bring benefits like recharging aquifers that provide drinking water and providing environments for complex riverside flora and fauna. Climate change may increase both floods and droughts. Rivers flood in many ways and understanding the characteristics of a watershed is critical before undertaking flood management projects. The National Flood Insurance Program (NFIP) offers a nonstructural way to deal with floods, but is inadequate. Neither levees nor NFIP requirements work with the natural processes of watersheds or value a variety of natural resources. More cost-effective, sustainable methods exist.

Rivers

What is a river?

It seems like a simple question, but there are many different answers. Is it the channel that usually carries water, which river pilots must know intimately? Does it include the floodplain, which carries water only periodically? What about groundwater that feeds into the channel and/or floodplain? Or is it the entire drainage basin that feeds water into the relatively small ribbon of blue we see on a map?

The most comprehensive answer is that a river includes the entire drainage basin, or watershed. Rivers start high and run downhill. Big rivers can start on mountaintops and run to the sea; small streams may run only a few miles before they end up in bigger streams. Because of the effect of gravity, everything that happens in a watershed can affect the river.

Rivers have four natural functions.

- *Transporting sediment*. Rivers start in upland areas, with steeper slopes in their upper stretches. Water moves rock, gravel, silt, and clay downslope. This is most important over longer periods and much of this work is done during floods.
- *Draining the watershed*. Rainfall and snowfall must be transpired by trees, evaporated back into the sky, or drained from the watershed by streams. This is most important over short periods.
- *Providing fresh water supply*. People, animals, and plants need fresh water to live and thrive. Streams and groundwater supply water, while riparian areas and wetlands can clean water of pollutants.
- *Transporting chemicals and nutrients*. Human life ultimately depends on the other animals and plants we use for food. If soil is depleted of nutrients, or if pollutants build up in soil, our food will not grow. Wildlife and farmed animals will move on or die. Food, recreation, and even spirituality can depend on a healthy web of food available to all creatures.

Besides these natural tasks of rivers, people use rivers and their floodplains for a number of other things.

- *Living*. People build in floodplains because the land is flat and the river offers many advantages.
- *Domestic and industrial uses*. We use water from rivers to irrigate, to provide industrial cooling, and to generate power, to name just a few things.
- *Diluting pollution*. Vast amounts of potentially dangerous substances around the world are disposed of into rivers, where they dilute and pose no problem. There is, of course, a limit to how much dilution a stream can afford before it too becomes polluted.
- *Navigation*. Many cities were founded on rivers because water provided the fastest, easiest, or cheapest line of transportation. Ports and navigability are still important assets to many communities and to worldwide commerce.
- *Freshwater fisheries*. River fishing is an important economic, recreational, and food source for many.
- *Recreation and tourism*. Fishing, boating, and hiking along streams are an important draw to many communities.

The important thing about rivers is not a list of what they're used for. When looking at floodplain or channel changes, it is imperative to remember that river functions rely on an interdependence of physical, chemical, biological, and societal processes. Defining a floodplain by law or practice, then focusing only on that narrow definition without taking into account existing wildlife and land-use practices, can give you a skewed picture of the river's health.

A complex biota of natural plants and animals is usually the most cost-effective way to protect water *quality*. The quality of water may be tainted because of pollutants entering the watershed even far away from the main stem of a river. However, some "pollution" actually results from eroding rock that includes minerals like phosphates that we don't want in our drinking, swimming, or recreational water. That's why you need to understand the characteristics of your watershed.

All these things can affect the amount and purity of water running in a river. The United States has 3.5 million miles of rivers. Of these, human development has substantially affected 81 percent, and 80,000 dams created reservoirs that impound another 17 percent. That leaves only 2 percent (about 70,000 miles) of our rivers in a relatively undisturbed state.

It is even easier to understand why watershed activities might reduce the *quantity* of water flowing down a river. Irrigation can take water from a stream channel and divert it entirely away, leaving a reduced amount of water in the river. Dams can block enough water that the stream can almost completely disappear below the dam.

Many places that regulate the amount of water taken from a stream do not ask how much groundwater you use. Groundwater is often an important source of recharging the surface water, and river flows can diminish if the use of groundwater is not in harmony with the river's need for replenishment.

During floods, channels can migrate, with a river staking out new ground. This is part of the reason levees are popular, because they seem to be able to hold a stream within its given banks. This often doesn't work, however, and almost never works long term without increased construction and upgrading of the levees.

Floodplains

The physical description and activities of the river and watershed are critical to understand. They influence the biologic and societal needs and potential uses of floodplains. As important as all the activities in a watershed are, in this book we are focusing on the floodplain. Human activity on floodplains can trigger quick and deadly responses from the river.

However, solutions to problems caused by floodplain development may be found in other parts of the watershed.

There are many definitions for a floodplain, but for our purposes, it's the land next to, or hydraulically or ecologically connected to, the flowing river. A one-year floodplain is the area the river covers in a flood, on average, once a year. The ten-year floodplain, then, is the area covered, on average, once in a decade. However, you can have ten-year floods in consecutive years, or with thirty years between them. While it would be extremely helpful to know when a ten- or twenty- or hundred-year flood would happen, weather happens when it does. An area might have five consecutive wet winters causing three ten-year floods, and then have nine drought years with zero ten-year floods.

Time

The basic floodplain used by the National Flood Insurance Program (NFIP) and in land-use planning is the 100-year floodplain, which is generally used in the United States as the standard to which we build. There are at least two problems with this.

First, delineating a 100-year floodplain (a flood that has a 1 percent chance of happening in a given year) is not easy and must be extrapolated from available data. In most cases, data come in a spectrum from precise river gages to inundation maps to oral histories. Even if you have 100 years of data, there is no guarantee that the largest flood in that time span was the 100-year flood. We have 160 years of data on the Mississippi. Was the 100-year flood the one in 1844? 1927? 1993? 2008? Or one that occurred 162 years ago or will happen three years in the future?

Few watersheds in the United States have the same population, land cover, and water use as 100 years ago. As populations increase, more buildings and roads are built, and more water is needed for home, office, and industrial use. All these things can change a floodplain. In its natural state, the main stem of the Mississippi flowed through gently sloped grass and forestland. That same land now contains Minneapolis–St. Paul, St. Louis, and New Orleans. Whatever the natural 100-year floodplain of the Mississippi was, development has changed it drastically. The same is true of other watersheds.

A study of the Johnson Creek basin in Portland, Oregon (often described as the last free-flowing stream in that metropolitan area), shows that the current average flow, after extensive urbanization of the basin, is equal to what would be about a seven-year flood in its natural state. A

100-year flood in its natural state is now a 25-year flood. These changes have occurred in less than 150 years, so the stream is still adjusting to its new regime.

A second factor in flooding is climate, which changes over time. As global temperatures rise, different areas of the United States will react in different ways. Precipitation patterns may change, watershed plant and animal communities may change, and rivers will have to adjust. Some places may have more flash floods. Others may have more floods in general. Still another result may be increased winter flooding and decreased summer flows.

Size

As one of the longest rivers with one of the largest watersheds in the world, the Mississippi offers lessons in looking at size.

The Mississippi is one large watershed, but it is also an interrelated set of smaller watersheds. For example, tiny streams along the east side of the Colorado Rockies coalesce into Cherry Creek, which flows through Denver. Cherry Creek flows into the South Platte River, which flows into the Platte River, which empties into the Missouri, and then into the Mississippi River. So ranching and farming activities, and a dam, in the uplands of Colorado influence the water quality and quantity of Cherry Creek, which then flows through Denver and is affected by that urban environment. In this way, the lives of the millions of people between the crest of the Rockies and the mouth of the Mississippi at the Gulf of Mexico ultimately affect the health and behavior of the Mississippi River.

Traditional levees have provided much of the flood management activity along the lower Mississippi (below Cairo, Illinois). Other strategies are available, like using setback levees and restoring wetlands, or rebuilding the barrier islands and wetlands that have been either deliberately or inadvertently destroyed in the Mississippi delta region.

Yet another strategy is to look at the smaller watersheds within the Mississippi basin and better manage water practices within each of those watersheds as part of a coordinated whole. The geology of these sub-basins differs across the country. Bedrock, slope, and soil types on the Rockies are very different from those on the Appalachians, or on the plains in between. River processes and flood management in an upland watershed with little soil development, steep slopes, and shallow streams are very different from those in a downriver watershed covering a nearly flat plain with hundreds of feet of soil and a deep, meandering river.

Floods

Rivers flood in different ways.

- Many rivers flood like the Mississippi: starting upstream, floodwaters slowly make their way downriver giving some places hours, days, or even weeks of warning. Properties along the river flood repeatedly. Dams and levees are a typical response.
- Especially in arid areas with steep slopes, or in urban areas, flash floods may predominate. They can also happen in other places that get inordinate amounts of rain. These floods happen when the ground cannot absorb intense precipitation. Water runs swiftly downhill into stream channels or dry gullies, causing a rapid increase in water height and ripping out trees, buildings, and bridges along the way. There is little warning that the flood is coming. Most flood deaths are due to flash floods, and about half those deaths involve people in cars.
- In areas with steep slopes and wet soils, mudslides are a part of the flooding process and cause more damage than the floodwater itself. The same precipitation that causes floods can saturate the soil, which then loses cohesion and slides downhill, mixing with water in stream channels. The resulting slurry of soil, water, trees, and rock can destroy houses, cars, or other obstacles in its path.
- Sinkhole flooding can be avoided if you understand the geology of the area. Karst topography, a bedrock generally made of limestone or dolomite, covers about a quarter of the United States. Groundwater dissolves the limestone, leaving sinkholes, sinking streams, and caves. Sometimes sinkholes and their floodplains are so large that they are not easily recognized. These sinkholes can also be masked by glacial depositions, as are common in the Midwest. During floods, water ponds in these depressions until it can drain through the subsurface conduits.

Flood Policies

The two major components of American flood policy are levees and the National Flood Insurance Program (NFIP). Both work on the assumption that you can accurately delineate a 100-year floodplain. Watershed development and changing climate patterns make that a shaky premise.

Discouraging the use of levees as a flood-control measure is a major theme of this book. The long-term costs in money, lost natural resources, and catastrophic failures are unsustainable. In the United States we began moving away from levees in the 1930s. However, we still have a system that includes local diking districts and the USACE that primarily focuses on levees.

The NFIP focuses on nonstructural solutions. Appendix A explains the history and practices of the NFIP, but a brief review is given here. Essentially the program allows anyone to purchase flood insurance. Most lending institutions require insurance for anything within a 100-year floodplain when financing involves the federal government and commerce. Relying on the certainty of a 100-year floodplain is an enormous drawback of this program. The emphasis given to that theoretic line leads most people to assume they are safe if not within that boundary, but increasing development can quickly render maps outdated. Other drawbacks to NFIP programs:

- Not taking into account what actions on one property do to properties downstream. They accept projects that simply move water faster through one reach and increase downstream flooding.
- Not looking at the effects of individual projects on the watershed as a whole.
- Not emphasizing the importance of upland storage or wetlands as potential flood mitigation processes.
- Not working with natural processes to manage floodwaters.
- Not requiring any analysis of future conditions of the watershed or future flood management needs.

We manage floods in a patchwork way, based largely on values set in the early twentieth century. There are better, less expensive, more cost-effective measures for flood management. They require, however, that the focus be on working with nature, not trying to control it.

Case Study: Soldiers Grove, Wisconsin

Soldiers Grove was incorporated in 1888 on the banks of the Kickapoo River in southwest Wisconsin. Soldiers Grove's population peaked in 1940 at 778. By 1975, 36 percent of the village's families earned less than $3,000 a year. As described in *Rebuilding for the Future: A Guide to Sustainable Redevelopment for Disaster-Affected Communities*, a 1994 Department of Energy report, the lumber industry, then farming and urbanization, began stripping vegetation from the river's watershed and the Kickapoo began to flood. The river hit Soldiers Grove with its first major flood in 1907. It dumped record amounts of water into the community again in 1912, 1917, and 1935. The 1935 flood finally made it clear to the villagers that flooding was a serious and permanent problem. In 1962, Congress approved a flood control plan for the Kickapoo River Valley—a dam and recreational lake near the top of the river, and a supplemental levee at Soldiers Grove, all to be built by the Corps. In the three decades that Soldiers Grove waited for dam construction to begin, the Kickapoo delivered four more floods, including a disastrous record inundation in 1951 that severely damaged downtown buildings and sped the community's economic decline. Construction of the dam and design work on the levee began in the late 1960s.

1. What Values or Assets Do You Want to Protect or Enhance?

Soldiers Grove was a long-settled community that needed to reduce its damages from floods to protect its economic and social health.

2. What Are the Apparent Risks or Opportunities for Enhancement?

Flooding was slowly destroying the town by threatening lives, property, and long-term economic stability. Its by-products (along with other events) were economic depression and the resulting loss of a younger generation who wanted more opportunities.

Specific problems for the town included the municipal well and location of the fire station. The well, located in the floodplain near the business district, became contaminated with even minor inundations. Their fire and rescue station was also located in the floodplain and could be cut off during floods when its services were most needed.

3. What Is the Range of Risk-Reduction or Opportunity-Enhancement Strategies Available?

The traditional answer, proposed by the U.S. Army Corps of Engineers, was a levee. The levee would cost $3.5 million to protect about $1 million worth of property. Each year for the next 100 years, the village would have to raise funds equal to twice its 1975 property tax levy to pay for maintenance of the flood-control structure.

Community leaders suggested that the federal government spend the same $3.5 million to help the community evacuate its floodplain and rebuild the business district on higher ground.

4. How Well Does Each Strategy Reduce the Risk or Enhance the Resource?

Levees would reduce flooding. Over time, however, they would need to be rebuilt higher to continue to protect the town.

Moving the village would remove the danger of flooding.

5. What Other Risks or Benefits Does Each Strategy Introduce?

The cost of maintaining levees is steep. If a flood did break through the levee, the results could be catastrophic.

The cost of moving an entire town is expensive, in terms of both money and intangibles. Removing the flood threat over the long term would pay for the short-term cost. Some of the history and sense of place would be lost by changing the location. However, new construction would allow for new building practices that create more energy-efficient structures, reducing some energy and environmental costs.

6. Are the Costs Imposed by Each Strategy Too High?

Levees are a proven commodity. Their risks and often exorbitant costs are known.

Moving a town is a much more experimental activity, with many unknown variables.

7. The Decision

With a small planning grant, Soldiers Grove hired a team of University of Wisconsin specialists to study the feasibility of relocation. The team concluded that relocation made sense.

In 1977, the village used its own funds to buy a site for the new downtown and begin the extension of utility services. The U.S. Department of Housing and Urban Development granted the village $900,000 in 1978 for relocation.

The villagers took the opportunity to add other benefits to their flood strategy. Soldiers Grove had become a community largely of elderly and low-income people. Completed in 1983, the relocation proved to be just what Soldiers Grove needed—a community heart transplant. Flooding was eliminated and there were other benefits. The downtown was moved back to U.S. Highway 61, bringing new economic life into the community. The old floodplain was developed into a municipal park, complete with tennis courts, picnic areas, and a number of other recreational features.

The extension of sewer and water services to the relocation site opened up new development area along its route. It also provided an opportunity for the village to fix a long-standing problem—discoloration and odor in the water in one nearby neighborhood, caused by aging pipes. Even more important, the village decided to protect its water supply by adding a second well and a larger storage reservoir outside the floodplain in the new downtown.

To improve public safety, Soldiers Grove also relocated the fire station to a site where emergency vehicles would not be disabled or cut off during floods.

They also decided to make all the new town center buildings energy-efficient and solar-heated, dubbing it Solar Village.

Conservation and renewable energy systems held the promise of increasing local self-reliance while insulating the community from oil and gas shortages of the future; reducing air pollution in a valley whose terrain often causes air inversions; and—most attractively—substantially lowering energy bills for businesses.

To capture these benefits, Soldiers Grove passed energy performance standards for its new buildings twice as stringent as those required by state law at the time; approved a community-wide "solar access" law that prohibited any building from blocking another building's sunlight; passed a law requiring that all new commercial buildings receive at least half their heating energy from the sun (the first such ordinance in the nation); conducted a careful analysis of the relocation site to determine its wind and sunfall patterns throughout the year; helped businesses position their buildings to make maximum use of summer breezes and solar heating while using earth berming and careful landscaping to block the buildings from winter winds; and conducted an evaluation of possible building materials to find those that were most economical, including costs to the environment. In the process, the village pioneered a new approach to disaster recovery.

There were some drawbacks. About a third of the original businesses closed before the move and never reopened. Not all the residents like the look of the new town, missing the charm of the original, traditional downtown area.

In August 2007, the biggest flood in its history hit Soldiers Grove. Floodwaters topped levees and high water stood for about ten days. However, because the town had moved uphill years before, there was little damage. The riverside campgrounds and park—built where the old

Main Street stood—flooded, but there was nothing to be seriously dam-aged. The Solar Village had no flooding.

Until 2003, the Crawford County Highway Shop was located in Gays Mills, 10 miles downriver from Soldiers Grove. During floods, when county services were most needed, the office and equipment could be flooded and unavailable. The 2007 flood inundated Gays Mills, but the Highway Shop, rebuilt on higher ground, was in no danger from floodwaters and county staff was better able to respond to the needs of the public.

Chapter 4

Natural Processes Must Drive Solutions

Change is a constant of rivers and floodplains. Understanding river processes is necessary to manage floodwaters. Rivers have physical, chemical, and biological characteristics. Climate affects rivers by producing the water and sediment that streams transport, and climate change portends major changes to some river systems. The biology of plant and animal systems has a major effect on water quantity and quality. Human activity throughout a watershed, especially on floodplains, can disrupt natural processes and make flooding worse. It is generally much less expensive to preserve natural processes than it is to re-create their lost functions. Working with, not against, a river's natural characteristics is generally the most cost-efficient and least disruptive course of action.

Natural Processes

Earth is a dynamic and ever-changing planet. Many of the resources that humans depend on—dry, flat land; clean water; fertile soils; productive fisheries; healthy forests—depend on the constant work of natural processes. To better understand how our actions affect these resources, we need to understand the physical, chemical, and biological processes that create and maintain our watersheds, rivers, and floodplains.

Weather and Climate Change

Energy from the sun is the driving force behind global climate patterns. Other natural processes that affect climate include wobbles in the Earth's orbit, ocean circulation and temperature patterns, volcanic eruptions, and upper atmosphere aerosols. Heat and energy from the sun evaporate water from the oceans, then wind carries this moisture over landmasses to fall as the rain and snow that fill streams and lakes.

The global patterns of high and low atmospheric pressure are altered at regional and local scales by mountain ranges and large lakes. The western slopes of the Cascade and Sierra Nevada mountains receive large amounts of precipitation from moisture-laden winds blowing off the Pacific Ocean. The eastern slopes of these ranges are in a rain shadow: most of the moisture has been removed from the air and little is left to fall. The western and eastern slopes of the Rocky Mountains exhibit the same pattern but usually with slightly less total annual precipitation.

Large lakes can also affect local weather. Many regions along the U.S. side of the Great Lakes receive large amounts of lake-effect snow. It comes from evaporation taking place as winds blow over the lakes, enriching the atmosphere with moisture.

Human activities can also affect climate patterns. Consensus is growing on the climate effects of the burning of fossil fuels and other practices that have greatly increased atmospheric amounts of carbon dioxide, methane, and nitric oxides. Predicted alterations from global climate change include rising sea levels, changes in snowpack and water storage, and changes in the location, timing, and intensity of storms.

Changes in precipitation patterns can change river flow patterns. Recent analyses of weather records show an increase in heavy rain events across the United States. Models of West Coast streamflow suggest higher winter flows and lower summer flows in the future. An increase in heavy rain events will likely lead to more flooding. This will increase the erosive power of the streams and rivers and lead to more transport and deposition of sediment in lower-slope areas of river basins.

Rivers as Corridors

Rivers transport water, sediment, nutrients, and energy from mountains to valleys and oceans. The physical, chemical, and biological setting of rivers affects this transport in ways that vary through time and space. Geological processes such as tectonic uplift and glacial advance and retreat formed the basic landscapes we see—mountains, valleys, and plains. Many landscapes are still responding to these forces.

Different terms are used to identify the area within which all surface water flows to a stream. In the United States, we use basin or watershed for this topographically defined landscape feature. The standard European word is catchment. Each watershed has a different geological, ecological, and human history. The particulars of each watershed and floodplain are unique, but the processes are not. Water, sediment, chemicals, and terrestrial and aquatic plants and animals are inseparable partners on this journey from sky to sea.

The tectonic action of raising rocks into hills and mountains is the first step in erosion. Over time, rock is broken into smaller and smaller pieces by mechanical and chemical processes known as weathering. In colder locations, freeze-thaw cycles can break up rocks. Glaciers and ice are powerful agents that can break off large pieces of rock or grind rock into flour-sized particles. Water and acids also break down rock by dissolving some of the binding material; these pieces are then transported by wind, water, and gravity to other locations.

Understanding erosion allows us to identify when sediment is exceeding natural rates and when it is within natural ranges. The physiographic and ecological setting of each stream sets these. Erosion, transportation of the eroded materials, and ultimate deposition of these materials are a fundamental function of watersheds.

Creation of Floodplains

Erosional particles move downhill primarily through stream corridors and their associated floodplains. Floodplains consist of zones, each with different ongoing processes that influence how they respond to human and natural disturbances. Figure 4.1 shows some of these zones.

Most early civilizations originated along steams on alluvial valley floors because of flat land, fertile soils, and ease of transportation. Many of our flood-control efforts take place in floodplains, trying to stabilize these areas and prevent the system from doing what is normal.

Active Terraces

Active terraces, as their name implies, are continuously changing. They are the lands the river covers with floodwater and wetlands. For planning purposes, the 100-year floodplain is often mapped. Active terraces are much broader in scope, including land that the river covers on a geologic timescale of thousands of years. River channels migrate across active terraces and floodwaters deposit sediment. Both processes wreak havoc on

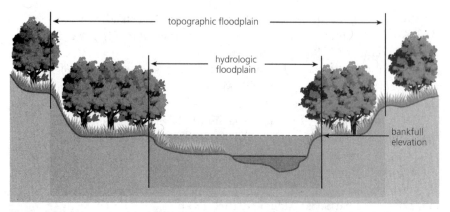

Figure 4.1
Inactive terraces, part of the topographic floodplain, are popular places for development because they often look stable, even though they can be at risk from floods or landslides. (Source: Federal Interagency Stream Restoration Working Group)

built structures. Ecologically, these areas are high in biodiversity, contain rich soils, and provide complex habitat for suites of animals and plants, both terrestrial and aquatic.

Inactive Terraces and Alluvial Fans

As a river flows through active terraces, the erosional energy of the water downcuts the channel into new material. Active terraces become inactive as new terraces are built at the new, lower level of the streambed. Several strata of inactive terraces, created through millennia, may stairstep up from the current stream channel.

Alluvial fans are fan-shaped sediment deposits at the mouth of drainage channels. They are created where a change in slope diminishes the energy available for sediment transport.

Inactive terraces and alluvial fans can give the impression of being permanent and unchanging, but they are not. Because they consist of materials that were previously transported by water, large precipitation events and changes in the active channel can result in sudden changes in these zones. Development often occurs in these areas because they are close to the river and look safe. They are, in fact, some of the most dangerous places to build because of threats from floods and landslides.

Bedrock Canyons

Rivers running through bedrock canyons typically do not have large active floodplains. The hardness of the bedrock protects the canyon walls from being eroded and restricts channel migrations. However, this very rigidity can result in extreme and unpredictable flows.

The Big Thompson Canyon in Colorado experienced a deadly and destructive flash flood in 1976 when a thunderstorm dumped more than a foot of rain in the watershed. Without a floodplain that the water could spread over, water levels rose rapidly and washed out campers, homes, and hikers. Within two hours, the flood had killed 145 people, destroyed 418 houses and damaged another 138, destroyed 152 businesses, and caused more than $40 million in damage.

Sediment-Rich Valleys

Rivers that flow through areas without confining bedrock canyons often occupy large valleys that are filled with deposited sediment called valley fill. Having been transported by water, these soils are unconsolidated and easily eroded. They are highly porous, meaning soil particles are interspersed with spaces that can be filled with air or water. In areas with high rainfall, these valley fills can store large amounts of water in the pore spaces.

This provides natural water storage that can be used by plants, be accessed by humans through wells, and provide constant water to streams, even during periods with no rainfall. The characteristics of these valley gravels also make them valuable for use in road construction and making concrete. However, the removal or compaction of these gravels decreases storage ability, increasing discharge.

These are very attractive locations for human development, including cities, farmland, and industrial complexes.

Floodplains

The natural function of rivers is to transport water, sediment, and organic materials from watersheds as part of ongoing hydrologic and sediment cycles. Water and sediment are both finite resources that are constantly being transported and stored throughout the world. Water's properties as a universal solvent mean that water is also transporting chemicals and nutrients as part of those recurring cycles.

While many of us think of surface water as the major component of the hydrologic cycle, the rivers and streams that we see every day are only a fraction of a percent of the freshwater that exists on Earth. As such, they are precious to our health and the health of the ecosystems upon which we depend. Misuse, whether intentional or not, can have severe and long-term consequences for us. We should pay as much attention to the condition of our rivers as we do our bank accounts. An assessment of our watershed and floodplain assets must include the benefits provided by functioning systems and the costs of altering those systems in both the short and long term.

Floodplains are very attractive places to build. They are typically level, close to water (which enhances their utility and their aesthetic value), and deceptively dry much of the time. Many of our flood-control efforts are to protect human structures in floodplains.

Floodplains as Systems

Climate, land cover, geology, topography, and land use interact to create river and floodplain systems. Watersheds act as filters and moderators of the delivery of water, sediment, and nutrients to rivers and oceans.

In smaller watersheds with narrow streams and modest streamflows, land-based processes tend to dominate stream processes. The presence or absence of large logs and boulders heavily influence stream characteristics. Roots and obstructions such as bedrock and boulders control the energy of the water to erode the streambanks. Much of the food used by aquatic organisms comes from leaves and other debris produced on the land.

In larger rivers with extensive floodplains, the river tends to dominate the land. Fast flowing, large quantities of water sweep trees into the river, and the river changes course across the floodplain with each major runoff event. Stream organisms get more of their food from aquatic production inside the stream, including algae, plankton, and macrophytes (large water-based plants).

Streamflow and flood patterns tend to be seasonal. During seasons of high precipitation or snow and glacier melt, flows respond to precipitation or melt by rising. During seasons without precipitation or snow or glacial melt, inputs from groundwater maintain the streamflows. The types of channels that exist are shaped by the quantity and source of water inputs as well as geology, soils, land use, and land cover.

Change is a constant of rivers and floodplains. Trees grow and die, stream levels rise and fall, erosion scours and fills. This dynamism is key to functioning aquatic and floodplain ecosystems. The unpredictable nature of when and where disturbances will occur creates a vibrant and complex physical habitat for plants and animals that are adapted to this environment. However, this same dynamism wreaks havoc on infrastructure and human-occupied areas that place a premium on stability and predictability. This is the heart of the struggle both to maintain natural processes and to make use of the services that rivers and floodplains provide to people.

Disturbances (changes) can be discrete or chronic. Most natural disturbances that affect watersheds and floodplains are discrete events. Floods, fires, windstorms, and volcanic eruptions are usually time- and space-delimited. They alter the physical environment and may also alter the interaction of populations and communities with each other and/or with the physical environment. Most natural disturbances alter the availability of resources for a time and then the ecosystem recovers. Some natural disturbances, like the glacial advance and retreat associated with climate variability, persist long enough to be chronic and alter the ecosystem structure and function for hundreds or thousands of years.

Disturbances are usually classified according to how often they occur (frequency), how long they last (duration), when they occur (timing), and how big they are (severity, intensity, or magnitude). For example, a high flow classified as a two-year flood has a 50 percent chance of occurring in any given year. It occurs relatively frequently, does not last very long, and is not very big. It usually does not significantly alter the overall shape or size of the stream or floodplain.

Lower frequency floods often cause greater change to riverbanks. A 100-year flood is a low-frequency, high-energy event. It may cut new channels, destroy sections of riparian vegetation and trees, and scour or deposit large amounts of sediment. However, if such floods can flow onto wide floodplains, the energy is broadly distributed.

River and floodplain organisms have evolved in conjunction with the disturbances present in riverine systems. They are adapted to disturbances within the natural range of variation in frequency, duration, and magnitude. Many riparian and floodplain trees, such as cottonwoods, need flooding for reproduction. Or, like willows, they are well adapted to the forces exerted by floods. Willows have flexible stems and the ability to root and grow from broken stems. When dams and other structures control river flows, the absence of natural disturbances can lead to changes in riparian vegetation because some species cannot reproduce without periodic floods.

Chronic disturbances may come from human alterations of the riverine system. Dams and diversions create new, long-term flow regimes that are different from what the stream organisms are used to and that affect their reproductive success and longevity. Changes in nutrient inputs from industrial or agricultural activities alter microbial communities and change water quality enough that new organisms become dominant. Disturbing the landscape cover—for example, changing from forests to cities—changes the timing, quality, and quantity of water delivery to streams. That, in turn, changes the physical form of the stream as well as the organisms that thrive in it.

Plants and Animals: Natural Flood-Reduction Tools

What do plants and animals have to do with flooding and river efficiency? Everything. Ecological processes—the interactions between living and nonliving components of the river system—should be the risk-reduction tools of first choice.

Flooding, as a hazard, is water flowing where it is not wanted, often transporting debris and pollutants with it. Rain has to flow through a labyrinth of biologic and physical structures to get to rivers and floodplains. It must flow through forests, fields, yards, roofs, and streets into creeks, streams, rivers, and on to floodplains to become a flood. Ecological processes can increase flood risks or diminish them.

Plants remove water through transpiration, store water in their tissue, hold soil, and strengthen and armor riverbanks. Animals assure healthy plant populations by pollinating flowers, fertilizing soils, distributing and burying seeds, tilling the soil, and eating the plants. Intact floodplains naturally provide a variety of ecosystem services such as flood regulation, water supply, and waste treatment.

These processes are not always intuitive. Several years ago in Yellowstone National Park, several streambanks that had been eroding began to stabilize. Willow, cottonwood, and alder trees began repopulating streamsides, helping to stabilize banks, shade and cool the water, and increase flows. Beavers returned, along with their dam building activities, further improving the habitat. Fish runs and bird populations increased, as have foxes and raptors. Ecologists at first could find no reason for these improvements. Creative thinking by scientists at Oregon State University found the answer: They were the result of wolves being reintroduced into the park. Before the presence of wolves, herds of elk spent most of their time staying close to drinking water and eating tree saplings and other riverside plant life. The introduction of a predator forced the elk to change their watering patterns. Staying relatively stationary made the elk more vulnerable to attack by wolves. As the elk were forced to forage farther from the streambanks, saplings were allowed to grow to adult trees. The presence of adult trees stabilized streambanks, which allowed other plants and animals to return to the ecosystem. Recovery took several years so the connection between stream conditions and wolves was not readily apparent. It took scientific detective work to tie the stream improvements to the introduction of wolves.

The Water Cycle

Water is a finite resource, or a closed cycle, as seen in Figure 4.2. Molecules of water vapor are continuously formed via evaporation or transpiration and circulated in the atmosphere, only to fall again to the ground or ocean. On average, this takes about eight days. Streams and rivers are part of this hydrologic cycle, but they contain less than 1 percent of our freshwater. The atmosphere contains about ten times more water than do rivers and streams. The soil mantle (not counting groundwater) of the Earth contains about fifty times more water than rivers and streams. Unfortunately, those are not easy sources of water for us to use.

Other important cycles constantly move material from one state or location to another, including the carbon, phosphorus, and nitrogen cycles. The nutrient cycles are an important part of the food webs, or trophic ecology, of rivers.

Most energy, usually accounted for as carbon, enters the aquatic system through photosynthesis. Some is produced by terrestrial vegetation and some is produced by algae and phytoplankton in the streams. The relative amounts of each vary depending on stream size, the type of vegetation and shading along the stream, and the availability of nutrients within the stream. The type of energy (food) available (for example, living algae or dead leaf litter) affects the type of organisms that consume the material. Some streams where fish return from the ocean to spawn derive energy inputs from ocean nutrients. This radiates throughout the system of consumers of the energy, which leads to a variety of predator–prey complexes that affect our recreational and commercial fisheries.

Figure 4.2
The water cycle includes precipitation, surface water, and groundwater. Land cover and climate change can significantly alter the water cycle in watersheds. (Source: Federal Interagency Stream Restoration Working Group)

A Four-Dimensional System

We used to think of streams as changing in one direction, from upstream to downstream. Now we understand rivers as a four-dimensional system, as shown in Figure 4.3. The dimensions are:

- *Longitudinal*—the length of the stream.
- *Lateral*—across the floodplain and valleys through which the river flows.
- *Vertical*—across which water and chemicals are exchanged between surface water and subsurface water.

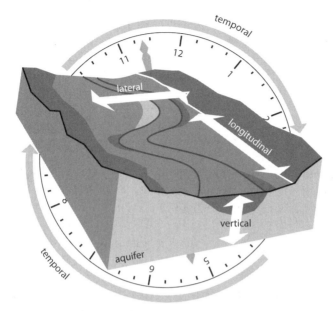

Figure 4.3
The four dimensions of a stream are longitudinal, lateral, vertical, and temporal. (Source: Federal Interagency Stream Restoration Working Group)

- *Temporal*—incorporates watershed history, as well as seasonal and annual variations in weather.

Most recently, these factors have been expanded to include the biodiversity that exists because of the physical features. We now recognize that altering any of these dimensions leads to a change in the plants and animals that can successfully inhabit the river system.

Connectivity between the three spatial dimensions is related to water levels in the river. Flooding is a necessary and important function of the river. It distributes nutrients, sediment, and seeds across the floodplain to support plants and animals. Connectivity of the four dimensions allows for material and information to be passed from area to area. Restricting floodwaters can dramatically change the plant and animal habitat of a floodplain.

Let's look at these four critical dimensions in more detail. The *longitudinal* dimension recognizes changes that occur from small headwater streams downstream to large rivers that empty into lakes and oceans. Erosion tends to be higher in the headwater regions because of steep slopes and channels. This eroded material is usually transported and deposited farther and farther downstream over time. High flow events have the energy to move the most material. Depending on where you are, the stream may be eroding, transporting, or depositing sediment.

The *lateral* dimension includes the importance of overbank flow or flooding processes to the structure and function of streams.

The *vertical* dimension reminds us that water exists not just in the stream channel, but also below it. Depending on the geology, there can be an extensive exchange of materials between surface and subsurface flows, both vertically and laterally. Groundwater is often neglected when assessing the resources of, and potential risks to, a river. There are also other zones, the most important of which is the hyporheic. Traditionally, water along river systems has been defined as either surface water or groundwater. However, other zones of water have recently been defined that are ecotones—or intermediate in nature (physically, chemically, and biologically)—between surface and groundwater. The *hyporheic zone* is a broad term that generally describes subsurface water that is composed of a mix of surface water and groundwater. It is an important part of the vertical dimension of streams. The dimensions of the hyporheic zone can vary dramatically depending on the stream size, stream discharge, alluvial porosity and volume, and vertical and lateral exchange rates. In small headwater streams, the hyporheic zone is typically on the order of several feet both vertically and horizontally or entirely absent due to lack of alluvium or significant exchanges of water. In large alluvial valleys, the hyporheic zone typically is on the scale of several yards below the streambed and up to 300 feet wide. In some highly conductive stream systems (e.g., Flathead River, Montana; Yakima River and Methow River, Washington), the hyporheic zone can extend more than 30 feet deep into alluvium and up to 2 miles wide across the floodplain.

It is easy to think of the stream as just the channel where we see water at any given time. However, the *temporal* aspects of streams show that the location and extent of water vary by season and year. Low floodplains are frequently flooded while higher floodplains flood only in extreme events. Long-term changes may be evident because of abandoned meanders or other features, but they may be impossible to discern looking at a current map. Historical changes in river channel, depth, or plant and animal habitats may need to be researched. The dimension of time also includes the future, incorporating expected changes in land use, urbanization, and climate change.

No site exists in isolation. All four dimensions—and the watershed as a whole—must be taken into account when seeking a strategy for addressing flood problems at any site. If any one dimension is disrupted there will likely be unintended consequences. If any part of the interrelated system is not working, it may need to be replaced with expensive design techniques.

Natural systems are resilient largely because of their complexity. A natural riparian corridor has a much more complex mixture of plants and animals than a grass-covered, engineered levee. The natural system can store more water (decreasing peak flows) and filter more water (leading to cleaner water without additional water treatment facilities) than a channelized system isolated from its floodplain.

Water Quality

Plants and animals are a critical, though often underappreciated, part of the watershed and especially of the floodplain. They depend on the water quality of streams. Destroying any one species of the biota can disrupt the system. If fish runs are important, for example, the water must not only be clean enough for them to live, but it must also be the right temperature. Temperature in small streams, in turn, is heavily dependent on streamside plant life.

Light

Light, provided by solar radiation, is fundamentally important to aquatic ecosystems. It is the ultimate source of energy used by organisms in the stream. The amount of light impinging on a river varies with latitude, season, time of day, altitude, atmospheric conditions, and riparian canopy cover. The amount of light energy reaching photosynthetic organisms in the river varies with water clarity, which affects light absorption and scattering. In small, clear, shallow streams, light reaches the bottom of the stream. In larger, murky, deep streams, light is rapidly attenuated (absorbed and scattered) and may not reach the bottom of the water column.

The amount of light is usually controlled by the amount of shading provided by vegetation along the river. Along large rivers, trees cannot shade the entire channel but smaller streams may be entirely shaded. This affects the source of organic material that drives the food web. In streams with more light, aquatic plants can fix carbon with photosynthesis, but in streams with low light levels, the carbon that drives the food web usually comes from terrestrial sources. Human alterations of the natural vegetation can change the source of carbon inputs to the stream, which can alter the entire food web.

Temperature

Stream temperature increases with absorption of solar radiation by the water column. There is usually, but not always, a strong relationship between air temperature and water temperature. Other factors that influence temperature include canopy and shading, groundwater inputs, elevation, latitude, channel depth and shape, and orientation (aspect). Temperature often increases when streamside vegetation is removed. In general, water temperatures tend to increase as one moves downstream, largely as a function of elevation and air temperature changes.

Climate change can affect vegetation. Drought can kill plants and trigger fires. Temperature change can cause species to change.

Temperature directly affects metabolic rates, growth, and reproduction in aquatic organisms. Temperatures above or below the range of acceptable temperatures can negatively affect species health and survival. Fish can be broadly categorized as preferring cold (less than 64 degrees Fahrenheit), cool (64 to 77 degrees), or warm (greater than 77 degrees) water. Cold-water fish include trout, salmon, and smelt. Cool water fish include muskellunge, crappie, and pike. Warm-water fish include bass, perch, carp, and catfish.

Chemistry

The chemistry of water is also important. In undeveloped watersheds, the underlying soils and geology exert a strong influence on the materials that dissolve in the water. Dissolved ions typically increase in disturbed watersheds. Suspended solids are often viewed as a sign of poor water quality. Some streams, however, such as glacial-melt streams, have naturally high levels of suspended solids and aquatic organisms are adapted to these higher levels.

Dissolved gases like oxygen and carbon dioxide play important roles in aquatic ecology. Dissolved oxygen (DO) is crucial for sustaining all aerobic aquatic life. Temperature, atmospheric pressure, and salt content of the water affect the solubility of oxygen in water and thus the maximum concentrations that can be present. Human activities that increase temperature and light and add nutrients can deplete DO and make streams unsuitable for fish or other animals.

Photosynthesis and respiration are biological processes that affect the oxygen content of water. In productive waters with many macrophytes, oxygen levels are elevated during the day due to photosynthesis and decreased at night due to respiration.

Groundwater is often low in DO and high in carbon dioxide (CO_2) due to the biological processing of organic matter in the soils that consumes oxygen and produces CO_2 and chemical oxidation. Organic pollution of streams from sewage can create a biological or chemical oxygen demand (BOD or COD) that can depress levels of DO. Concentrations of oxygen vary around the country as a function of temperature, salinity, and in-stream oxygen demand. Introduced pollutants, such as bacteria, herbicides, pesticides, metals, and oil also contribute to water quality conditions.

Water quality affects aquatic ecology in various ways. Light is necessary for photosynthesis and also is correlated with the movements and feeding patterns of organisms from zooplankton to fish. Turbidity from either organic or inorganic sources can increase water temperatures because suspended particles absorb more heat. Higher temperature leads to lower levels of saturated DO. Low levels of DO can limit the types of organisms that can survive and thrive in the stream. Higher turbidity also reduces the amount of light penetrating the water, which reduces photosynthesis and the production of DO.

All organisms have specific temperature and chemical ranges in which they can survive. The concentration of total dissolved solids affects the water balance in the cells of aquatic organisms. Water with pH less than 6.5 or greater than 8.0 stresses the physiology of most organisms and can reduce reproduction. Low pH can mobilize toxic elements and make them available for uptake by aquatic plants and animals. This can produce conditions that are toxic to aquatic life.

Suspended solids can serve as carriers of toxics, which readily cling to suspended particles. Nitrates in excess amounts can accelerate eutrophication, causing dramatic increases in aquatic plant growth and changes in the types of plants and animals that live in the stream. This, in turn, affects dissolved oxygen, temperature, and other characteristics. Phosphorus in excess amounts can also lead to accelerated plant growth, algae blooms, low dissolved oxygen, and the death of certain fish, invertebrates, and other aquatic animals. Bacterial concentrations, as identified by *E. coli* counts, not only pose human health risks but also cause cloudy water, unpleasant odors, and an increased oxygen demand. Toxic metals and organic compounds can bioaccumulate in the aquatic system and lead to human health hazards and reproductive issues in aquatic organisms.

Resistance and Resilience

Resistance and resilience are important concepts. Systems that are resistant to disturbance do not change their structure and function much when disturbed. Physical and biological characteristics may shift slightly but maintain their original relationships. A system that is not resistant may change dramatically after disturbance, but if it is resilient, it will recover to its approximate original state once the disturbance is over or removed.

Complex, self-organizing systems tend to be resilient and can either resist or recover from disturbances. Extinction of a keystone species like beaver or the introduction of a species that alters light and energy like the zebra and quagga mussels can cause long-term changes. If the disturbance can be ended, recovery may take place. However, there is usually an unidentified threshold that, once crossed, makes it impossible for the system to recover to its original condition.

Conclusions

The major effect of most human activities in the watershed is to change the naturally occurring land cover. Agriculture, forestry, and urbanization all transform native forests or prairies to highly modified environments. Human activity in the watershed and floodplain often leads to unpredictable and large-scale disturbances and chronic changes.

Well-vegetated soil rich in organic matter allows more water to sink into the soil, delaying and dampening floods. The roots of vegetation along riverbanks help hold the soil in place and reduce erosion. Humans have created major chronic disturbances with dams, water diversions, channelization structures, and urbanization along rivers and floodplains.

Replacing vegetation with human infrastructure that compacts or paves the ground surface leads to more and faster runoff of precipitation. That creates larger and more frequent floods, causes more erosion and sedimentation, and carries more pollutants into the rivers and streams. This leads to human efforts to control the river with dikes, dams, and other structures that further focus and concentrate the flows, which further increases flood depths and damage. Estimates from Illinois place the dollar value of floodplain storage of floodwaters as high as $3,200 per acre per year.

Our knowledge of how these systems work has improved over time as we have learned more about the physical and biological processes of rivers and floodplains. Gravity provides energy to stream systems and this energy needs to be dissipated to reduce extensive scouring and downcutting. Energy is dissipated by bends and meanders that increase the flow distance per drop in elevation. Energy is also absorbed by blockages such as boulders, logs, and beaver dams.

Historically the approach to flooding was to straighten channels and remove natural blockages. The idea was to transport water through the system as fast as possible. These actions increase the energy of the river, which can then create further disturbances. This can lead to actions akin to an arms race where more disturbances from flood lead to more alterations of the floodplain in terms of straightening, narrowing, and confining the river, which leads to higher and more energetic river flows. Civil engineer Charles Ellet reported to Congress in 1852 that:

> The extension of the levees along the borders of the Mississippi, and of its tributaries and outlets, by means of which the water that was formerly allowed to spread over many thousand square miles of low lands is becoming more and more confined to the immediate channel of the river, and is therefore, compelled to rise higher and flow faster, until, under the increased power of the current, it may have time to excavate a wider and deeper trench to give vent to the increased volume which it conveys. . . . Future floods throughout the length and breadth of the delta, and along the great streams tributary to the Mississippi, are destined to rise higher and higher, as society spreads over the upper States, as population adjacent to the river increases, and the inundated low lands appreciate in value.

This prescient comment acknowledges the natural physical processes that control the transport of water and sediment through a watershed. We now also know that the natural disturbance regime creates a dynamic and vibrant set of habitats for plants and animals. Where we degrade these processes, we must replace them—usually at considerable cost.

Case Study: New York, New York

The Catskill/Delaware watershed provides 90 percent of the drinking water to New York City (NYC). The 1,600-square-mile watershed allows excellent natural filtration for the water, bolstered by chlorination to kill microorganisms. However, increased development and impervious surfaces, agricultural runoff, and discharges from wastewater treatment plants threatened the purity of the water. The same factors that tend to worsen flooding also worsen water quality and the recipe to improve both is often the same.

1. What Values or Assets Do You Want to Protect or Enhance?

There was a need for 1.3 billion gallons per day of clean drinking water to continue to be delivered to the 9 million people of New York City.

2. What Are the Apparent Risks or Opportunities for Enhancement?

Water came from the Catskill Mountains, 125 miles away from the city. Natural processes in the Catskills kept the water clean. Roots and microbes broke down contaminants, plants absorbed nutrients from fertilizer and manure, and wetlands filtered nutrients and drew heavy metals out of the downstream watershed. Development in the Catskills threatened these processes.

3. What Is the Range of Risk-Reduction or Opportunity-Enhancement Strategies Available?

Something had to be done, or the EPA would require new filtration systems. The new filtration systems would have an initial cost of about $7 billion and annual operating costs up to $400 million.

Another option was to restore the health of the Catskills' ecosystem.

4. How Well Does Each Strategy Reduce the Risk or Enhance the Resource?

Water treatment systems are commonplace in the United States and work very well.

The natural Catskill filtration has worked for NYC for decades and can continue to work in the future with some improvements. Upgrading this system provides other benefits as well:

- Preserving open space and the rural character of the Catskill/Delaware watershed, increasing opportunities for agriculture and forest-based businesses.
- Providing an opportunity for market development of previously unrecognized economic opportunities, particularly tourism. This is a new and growing field in ecosystem management.
- Protecting other valuable services such as flood control for Catskill basin residents and storage of carbon by plants.

5. What Other Risks or Benefits Does Each Strategy Introduce?

The risk for the new filtration system is primarily that of major costs throughout the lifetime of the system. There is also the possibility of breakdowns or a terrorist attack that would allow hazards to enter the water system of a major city.

The risk for the Catskill filtration is continued dependence on a sole source of cleansing for the city's water supply. Moreover, the city does not completely control the source and there are significant transport issues.

6. Are the Costs Imposed by Each Strategy Too High?

The costs of maintaining a built system far exceed the costs of maintaining the natural environment.

Columbia ecological economist Geoffrey Heal, quoted in a 2005 *National Wildlife Magazine* article, explains: "The traditional argument for environmental conservation had been essentially aesthetic or ethical. It was beautiful or a moral responsibility. But there are powerful economic reasons for keeping things intact as well."

New York City, however, would then be in the position of controlling a watershed that is not within its boundaries. That could lead to political irritation and disenfranchisement for those who live within the watershed. NYC must rely on its political power to keep control of the watershed within its authority.

7. The Decision

NYC will continue to use the Catskill/Delaware watershed and upgrade it as necessary. The city is implementing extensive watershed management measures, including water quality monitoring and disease surveillance, land acquisition and comprehensive planning, and upgrading wastewater treatment plants.

The foundation of the watershed protection program is land acquisition, which addresses the potential political friction between NYC and residents of the watershed. The city is more than one third of the way through a fifteen-year program. Between 1997 and 2003, NYC obtained or had under contract more than 52,000 acres at a cost of $131 million. More than 70 percent of the acreage obtained is in high-priority areas, including 1,200 acres of wetlands. NYC is also working with local land trusts to increase solicitation.

Not only will this provide NYC with water, but it will also improve conditions in the watershed for the residents there. Potential benefits for them include support for their own water quality, flood management, and tourism opportunities. Most of the population lives near streams in the watershed. A variety of organizations offer technical and/or financial assistance to watershed residents for flood protection, habitat enhancement, septic and sewage systems, and stormwater controls.

Managing many small streams protects the natural characteristics in the larger watershed, preserving its natural ability to absorb precipitation and avoid floods. At the same time, residents in the watershed have an opportunity to continue the limited agricultural, urban, and recreational activities that supply their economic base. The unspoiled character of the area provides an opportunity for tourism. By designating the area as a park and forest preserve, the area will continue to benefit New York City and local residents.

Chapter 5

Our Relationship to Rivers

On one hand, we adhere to a romantic vision of rivers essentially natural in character; rivers embody aesthetic ideals of nature and wilderness; rivers have provided a stage for exploration, pioneered settlement, and the advance of civilization; rivers give identity and sense of place to hundreds of communities. However, we have also exploited the short-term economic values of rivers as conduits for waste, as sites of cheap, easily developed floodplain land, or of investment in rapidly appreciating riverfront view property, while tending to discount long-term or externalized costs associated with flood hazards, erosion, pollution, and habitat destruction. Our challenge is to understand and work with the relationships between values that allow both short-term uses and environmental sustainability.

The Value of Rivers and Floodplains

People have long recognized the value that rivers and floodplains provide. From simple early settlements to complex civilizations, many human establishments are located in fertile floodplain valleys. The land was flat and good for agriculture; the river provided transport of people and goods, and diluted wastes; and fisheries provided reliable sources of food. Today we have other modes of transport and can supplement croplands far from floodplains with manufactured fertilizers and soil amendments. But watersheds, rivers, and floodplains are still vitally important to humans.

Rivers and ecologically intact watersheds and floodplains provide enormous economic, social, and cultural benefits to society. A study by the state of Washington Department of Natural Resources gives us one monetary figure. Between 1979 and 1989, 427,000 acres of primary forest in the state were converted to other uses: 48 percent to urban expansion, 28 percent rights-of-way, and 24 percent agricultural uses. It would take at least $2.4 billion to build a stormwater system equivalent to what that forestland provided in clean water and flood control.

Many of the assets we wish to protect and enhance relate directly to established uses of rivers and river environments. In many cases, the physical infrastructure to support those uses has significantly modified the rivers themselves. We have become increasingly aware of the negative effects of these changes that frequently lead to conflicts and trade-offs among values.

However, in many cases, these are relatively short-term conflicts related to the life of infrastructure because it was not designed to work with the river's natural processes. This book suggests that people can continue to use streams, but better designing and planning can reduce the effects we have on rivers and floodplains. We can protect and even enhance the natural physical and biological assets that rivers offer.

Hunter-Gatherer Societies

Since the dawn of our species, rivers and their extended environments have provided an important place for human life. Economic activities of early societies drew upon the natural resources of a river—its water, riparian borders and floodplain, and flora and fauna.

Many hunter-gatherer cultures, some of which have persisted into modern times, developed patterns of settlement and economic activity attuned to the natural, seasonal cycles of a river's flow and its aquatic and terrestrial life.

Athabaskan cultures, for example, spread through interior Alaska and large areas of northern Canada. They developed a limited pattern of nomadic living in which summers were spent in fishing camps along rivers to take advantage of summer and early fall fish runs. They invented technology, including nets and fish wheels, to harvest salmon and other fish directly from campsites on the riverbank.

Each year, temporary dwellings were reconstructed from light timber, sticks, and thatch. These were not expected to survive the next spring flooding. Winter settlements with more substantial, permanent dwellings were established on higher ground set back from rivers, safe from floods. Using boats in the summer and dogsleds in the winter, rivers provided highways to access natural resources (winter hunting, for example), and for trade and to establish relationships with other communities.

How hunter-gatherer societies affected natural systems is a subject of current research. There were effects, certainly, though extremely limited in comparison with those of even a modern agrarian society. Hunter-gatherer societies did use fire as a means of clearing selected areas of forest or shrubland to improve habitat for wild grazing animals, or in some cases as a means of herding animals during hunting. Some floodplain areas in the Pacific Northwest, including parts of the Willamette Valley in Oregon, were originally forested but later cleared through fire practices of Native Americans, and were not natural grasslands as initially interpreted.

Using wood for fuel and harvesting fish and other food sources was in most cases so limited that it had only a very small effect on the natural systems producing these resources. An exception to this in certain places around the world may have been the contribution of humans to the extinction of megafauna. These appear to have occurred concurrently with human migrations into Australia and North and South America. In North America, many species of large mammals disappeared during or shortly after the retreat of the last continental ice sheets between 10,000 and 12,000 years ago. Many Neolithic sites at widely scattered locations are also dated to this time. Though the human role in these extinctions is unproven, the disappearance of megafauna such as species of large beavers undoubtedly had an effect on the subsequent evolution of river systems and their floodplains.

Early Agrarian Societies to Urban-Agricultural Civilization

Civilization, as we define it, began on floodplains. The first extensive cultivation of land and the first permanent agricultural communities were along rivers and their floodplains. Valleys of the Tigris-Euphrates, Nile, and Indus rivers provided the environmental conditions and natural resources for settlement perhaps as early as 8,000 years ago.

The settlement of formerly nomadic peoples in these valleys, their concentration into communities along rivers, and domestication of grains, other crops, and animals, was a gradual process. What emerged in these locales, roughly 6,000 years ago, were forms of social organization we refer to as "civilization."

To use the term of historian Donald Worster in his 1985 book, *Rivers of Empire*, hydraulic societies developed in these valleys. They were organized around the manipulation of water and floodplains in an otherwise arid environment—irrigation and drainage; modifying riverbanks; building levees, dikes, ditches, and canals; leveling fields and draining or filling of wetlands; cultivating nonindigenous plants as crops; and husbandry of domesticated animals.

A hydraulic agricultural economy on a large scale required an increasing measure of centralized political control, planning, and engineering, supported by taxation and based on a system of written land records. This social and political organization led to emergence of the first cities, such as Ur on the Euphrates. This linkage between agricultural and urban occupancy of floodplains has persisted through history. This pattern of cities developing in floodplain agricultural regions was replicated throughout Europe and parts of North America.

The Colorado River is sometimes referred to as the American Nile. In its lower reaches, like the Nile, it flows through an extreme desert, conveying water from distant mountains to a region of almost no rainfall. The lower Colorado River and adjacent areas of Arizona, Nevada, and southern California are perhaps our foremost example of a modern hydraulic society based on massive works of water storage and distribution for irrigated agriculture, industrial, and domestic uses.

The sustainability of these systems is questionable today, given the limitations of water supply in relation to burgeoning demand, climate change, and an array of ecological impacts. Impoundments and changes in flow regime, evaporative losses from reservoirs, and direct water withdrawals have resulted in habitat loss, advance of invasive species in riparian zones, and disappearance of the wetlands of the Colorado River Delta. Irrigated agriculture has produced its own set of effects, including runoff polluted by agrochemicals and soil salinization.

Hydraulic societies represent an extreme case in human use of rivers and floodplains, and a highly problematic one for the regions where they have been established.

Eighteenth and Early Nineteenth Centuries River Valley Settlement in America

During westward expansion from the late eighteenth to the mid-nineteenth century, the development of major river cities in the interior of the United States was largely driven by using rivers as the primary transportation network. Cities and towns developed on the Ohio, Mississippi, Missouri, Tennessee, Cumberland, Alabama, Arkansas, Red, Rio Grande, Sacramento, Columbia, and Willamette as well as many smaller rivers, because the rivers provided transportation from agricultural communities to national markets.

On eastern coastal rivers, many towns were established on the Fall Line, a 900-mile-long escarpment where the Piedmont of the Appalachians descends steeply to the Atlantic coastal plain, often producing rapids or waterfalls. Cities on this line include Trenton, New Jersey, Washington, D.C., and Columbus, Georgia. Waterpower from the falls was used first for milling grain and later for other industrial purposes. In America, unlike the examples from antiquity, river towns were often settled first, then agriculture developed, which in turn fostered the growth of these towns into centers of commerce, and later, of industry.

The Mississippi river system alone encompassed more than fifty mainstem and tributary rivers once navigable by steam vessels for a total distance estimated at 10,000 to 16,000 miles. During the nineteenth century, some 8,000 steamboats operated on this system, and steam navigation played a central role in the agricultural settlement and urbanization of the entire central United States.

During the age of river transportation, riverfronts were the focus of urban life and economic activity. A daguerreotype photo of the Cincinnati riverfront in 1848 (part of which is seen in Figure 5.1) shows a 3-mile stretch of public landing along the Ohio River lined with warehouses, stores, shipping and trading companies, shipyards and marine

Figure 5.1
The Cincinnati riverfront in 1848 shows the close development to the river, lack of amenities, and the importance of steamboats. (Source: The Collection of The Public Library of Cincinnati and Hamilton County)

service industries, hotels, amusement, and other businesses that consti-
tuted the city's original downtown. Fifty-two steamboats are evident, of
more than 4,000 vessel arrivals that occurred there that year.

By the late nineteenth century, railroads had become the dominant
mode of transportation. Yet the urban centers established by river transport remained and remain today, as does continuing urbanization around
these cities and on their floodplains.

Nineteenth- to Mid-Twentieth-Century Agriculture, Transportation, and Cities

Railroads were the first of many influences that caused cities to turn their
backs on their rivers. This occurred in stages. Many early rail lines paral-
leled rivers. Rail yards and depots were built in the level land of floodplains or along riverfronts where connection could be made to river
transport. Over time, railroads brought about a reorganization of the nation's urban hierarchy and transportation network. Major rail routes became independent of rivers but tended to cross them at established river
cities where markets could be tapped, especially when cities contributed
to railroad bridge construction. Second-generation rail terminals were
built away from the river, often with a relationship to second-generation
downtowns that sought higher, less flood-prone land set back from the
river.

Highways produced a new scale and pattern of urban expansion
largely unrelated to the river, except to the extent that level floodplain
land was attractive because of lower up-front development costs.

Mainstem river traffic did not completely disappear, but evolved to
barge transport of bulk commodities such as coal and grain that required
only a few specialized port facilities along the length of a river. Most existing urban riverfronts were bypassed. Riverfront districts declined into
marginal uses and disuse, with property owners often unable to cope
with the cost of recurrent flood damage. Construction and periodic raising of levees and floodwalls provided some protection from more frequent moderate flood events, but separated cities functionally and visu-
ally from their rivers.

Ironically, the agricultural settlement fostered by river transportation
gradually contributed to its demise. The wholesale clearing of Midwest
forests for agriculture is an example. Ohio, 98 percent forested at the
time of the American Revolution, had only 7 percent forest cover in

1900. Deforestation coupled with the channelization of streams led to both more frequent severe floods and reduced summer flows. Consequently, navigability was lost in all but the largest tributaries of the Mississippi system. This happened despite efforts at river improvement through snag removal, dredging, bank stabilization, and introduction of slack water systems of dams and locks on smaller tributaries.

Agricultural expansion resulted in soil loss and sedimentation of rivers. The glacially deposited till plains of the upper Midwest had some of the deepest topsoil horizons anywhere in the world. Early agricultural practices were not conducive to soil conservation, and the introduction of mechanization such as Cyrus McCormick's harvester in the 1850s only accelerated agricultural expansion and soil loss. Soil conservation practices were not widely employed until the early to mid-twentieth century. By that time, sediment deposition had decreased the channel cross section of many streams in the Mississippi Valley, contributing to increased flooding and loss of navigability.

Mining also contributed to the sedimentation of rivers. The most notable nineteenth-century example was hydraulic mining for gold in the headwaters of streams draining the west slope of the Sierra Nevada in California. Powerful jets of water propelled by steam pumps washed gravels down from steep riverbanks and terraces. The millions of tons of sediment that washed down destroyed habitat and fisheries, caused floods, threatened navigation on the Sacramento and San Joaquin rivers, and affected areas as far downstream as the San Francisco Bay. The California legislature eventually banned hydraulic mining.

With the massive industrialization that occurred in most cities from the 1880s through the 1920s, the major role of rivers became that of sewers for industrial and domestic wastes. At the same time, most river cities still relied on their rivers for potable water supply. Primary treatment of urban sewage did not become common practice until the late 1950s. Secondary or higher levels of treatment to national water quality standards awaited the Environmental Protection Agency (EPA), the Clean Water Act, and subsidy programs that began in the 1970s.

Mid-Twentieth Century to Present, Urban Riverfronts and Floodplain Urbanization

The federally financed, so-called urban renewal programs of the late 1940s through the early 1970s led to the wholesale clearing of riverfront districts in many cities. Many historical, cultural, and architectural re-sources were unappreciated at the time and lost forever. The first gener-ation of redevelopment in these areas was often unrelated to the river, yet massive in terms of environmental and visual impacts. Highways, including at-grade, elevated, or below-grade expressways, often inte-grated with levees or floodwalls, were the final step in structurally sepa-rating a city from its river. Development imposed hard edges and ex-tensive impervious surfaces along river shorelines. Sports facilities with attendant parking lots or structures covering tens to hundreds of acres were a common feature of this first generation of riverfront redevelopment.

Since the 1970s, and especially since 1990, we have seen a reawaken-ing to the historic, cultural, and amenity values of rivers. A second, and in some cases a third, generation of riverfront redevelopment is trying to recapture those values.

As their riverfronts evolved, cities have also shared in the urban ex-pansion that has typified U.S. metropolitan areas since World War II. In the Midwest, many river cities have seen substantial decline in central city population since the 1960s. At the same time, most metropolitan areas have continued to grow, in many cases at least doubling in popula-tion since that decade.

The extent of urbanized land area has increased by a much larger fac-tor, and has involved extensive development on floodplains and adjacent land. By the mid-1980s, in a number of U.S. metropolitan areas, 20 per-cent or more of the urbanized area was in floodplains. Examples include Charleston, South Carolina, with 40 percent of the urban area in flood-plain; Fargo, North Dakota, also with 40 percent; Dallas, Texas, with 22 percent; Omaha, Nebraska, with 33 percent; St. Louis, Missouri, with 30 percent; San Jose, California, with 29 percent; and Monroe, Louisiana, with 81 percent of its urban area in floodplain. In the case of large cities such as Dallas and St. Louis, the total extent of floodplain urban area was immense—about 140 square miles in each case.

Urban expansion in floodplains is not limited to the older river cities and their environs. Younger, high-growth metropolitan areas in the Far West, including cities not historically connected to major rivers, also have had increasing problems with floodplain development. Examples include development on or near dry washes, arroyos, or alluvial fans sub-ject to flash floods or debris flows in the Southwest. In many parts of California and the Pacific Northwest, the primary concern is urban ex-pansion into formerly rural valleys.

Cincinnati, Ohio

Cincinnati provides an example of each of these phases of riverfront evolution. It was originally platted in January 1789, on the Ohio River. A year later, its population was nearly 700 people, which grew to 10,000 by 1820. Cincinnati was becoming the Queen City of the West.

As with many nineteenth-century cities, Cincinnati depended on the river, seen in Figure 5.1. Settlers traveled to their new homes in the Midwest on the Ohio River. Riverfront businesses at that time included hotels, restaurants, and taverns to meet their needs. Farmers used the river to send crops down the Ohio and Mississippi rivers. Steamboats were manufactured in town. In the early 1800s, Cincinnati became an important meatpacking center, and was once known as the Porkopolis of the United States.

By the late 1880s, Cincinnati had almost 300,000 people, and the densest population of any city in the United States, with an average of 37,143 people per square mile. In 2000, its population was 365,000, having remained nearly steady for over a hundred years.

As with other cities of the time, the commercial success of the city meant catastrophe for the river. When city planning and zoning began in 1925, the riverfront district was called unrecoverable. The 1937 flood, worst in the city's history, accelerated the district's decline, and a 1939 redevelopment plan called for clearing all buildings and removing streets for a distance of six blocks back from the river along 2 miles of riverfront, the entire lower terrace of the city. The plan envisioned a four-lane parkway and baseball park, retaining the public landing, and leaving much of the area in parks and open space. By the time the clearance took place in the early 1960s with federal urban renewal and highway funding, the parkway was expanded to an eight-lane, below-grade Interstate freeway segment, effectively separating the downtown from the river.

The stadium, completed in 1970, became a huge multipurpose sports facility built on top of what had been the historic public landing. Riverfront Stadium and its parking structure incorporated its own floodwall and an elaborate internal flood-proofing system, making it one of the most expensive such facilities built up to that time. Parking lots and an indoor coliseum completed the development, effectively covering the entire central riverfront with an impervious concrete surface and hard shoreline edge.

Almost as soon as this redevelopment was completed, there was a realization that important community values related to the river had been sacrificed. In the mid-1970s, as a bicentennial project, more than a mile of riverfront upstream of the redeveloped area was acquired by the city. It was developed for passive and active recreation, with walkways, view-points, interpretive signs, and displays related to the natural and human history of the Ohio River. Although this linear park incorporated con-siderable green space, it was decided not to re-create a more natural shore-line. Instead, a major structural feature, the Serpentine Wall was con-structed, a sinuous line of steps extending along and down into the river. The stepped wall allows access to the river at all river stages for fishing, kayaking, and other activities. With continued river cleanup, even swim-ming will be possible. A later extension of this park upstream included restoring a vegetated shoreline and incorporating flood-tolerant recre-ation facilities and public art.

These efforts did not satisfy demands to redress the ill-conceived re-development of the central riverfront. In 1999, Cincinnati embarked on one of the largest redevelopment projects in U.S. history. It included re-moving Riverfront Stadium and constructing new sports facilities on ei-ther side of the central riverfront area. The riverfront itself will be de-voted to parks, museums, a reconstructed public landing, and restoration of about a half-mile of natural shoreline. This will allow the river to re-connect with part of its floodplain.

The new baseball park at one end of the development has a much smaller footprint than the previous stadium and incorporates flood-tolerant designs that are less expensive to build. Lidding the below-grade freeway will enable the riverfront to reconnect to the city. On the lid, a mixed-use neighborhood is being developed with design and architec-ture that reflect the historic context of the original riverfront district. Figure 5.2 is an artist's drawing of the proposed development.

Admittedly, this project depended in part on the investment made in a new generation of sports facilities. However, the fact that a medium-sized American city would invest hundreds of millions of additional dol-lars to reestablish its connection to a river speaks to the powerful values associated with rivers.

Human Activity and Biological Change

Human activity can significantly alter vegetative cover. Logging seri-ously alters forest cover, but if desired, forests can be replanted. Road building alters surface conditions and roads often act as extensions of the channel network carrying water and materials directly to the river. Ri-parian and floodplain vegetation is often lost during farming. Fire sup-

Figure 5.2
The future riverfront of Cincinnati will look very different, with a mix of uses and a half-mile of restored shoreline. This major redevelopment project shows how important riverfronts are to cities and their residents. (Source: Port of Greater Cincinnati Development Authority)

pression in forests causes species changes. Agriculture and urbanization can radically and often permanently alter watershed cover. Pumping water from aquifers can alter the water table and change vegetation.

Monocultural agriculture tends to strip nutrients and minerals from the soil and send them to population centers where they ultimately end up in sewage systems that discharge into water bodies. Logging carries nutrients and minerals from forests to population centers where the wood is ultimately burned or buried in landfills. Mining shifts minerals to population centers where they can become pollutants that foul surface waters. Trans-basin transfers of water cause both depletion and surplus in the areas affected.

These transfers have questionable effects on long-term sustainability. Too often these nutrients and minerals end up in water bodies where they can cause secondary problems. Oxygen can be depleted by algae blooms, killing fish. Toxic metals can accumulate in aquatic species and disrupt food chains and limit human consumption. The dead zone that forms off the mouth of the Mississippi (an area that varies in size, but can cover up to 8,000 square miles) is an example of excess nutrient loading that causes algae blooms and fish kills.

Biological disturbances can also occur and affect the physical and biological features of the river and floodplain. One of the largest human-caused biological disturbances was the decimation of beavers from most of the streams in the United States. Extensive trapping and hunting reduced the population greatly and in many places totally extirpated this species. As harvesters of trees and dam builders, beavers exerted strong local control on smaller streams in watersheds. Effects of beaver dams include raising water tables upstream of the dam, reduced water velocities, sediment storage, increased water storage, improved water quality, more waterfowl nesting sites, shifts in aquatic invertebrate community structure, increased amphibian habitat, and changes in riparian vegetation. It is possible that streams today without beaver populations have a very different set of physical and biological characteristics than they had historically. Some states (e.g., Washington, Oregon, and Wyoming) are now reintroducing beavers to reinstate these stream and riparian characteristics.

The other major biological change (again human-induced) that can severely alter natural processes in streams and watersheds is the introduction of exotic species. It was once common practice to introduce game fish into lakes and rivers for anglers. Many of these fish were predators and ate other native fish or competed with native fish for invertebrate prey. There is evidence that nonnative fish are contributing to native amphibian declines in California. European mussels have invaded many lakes and rivers in the United States. They are filter feeders and can filter huge volumes of water compared to native species. This tends to remove turbidity from the water and allows light to penetrate to much greater depths. This changes where and which kinds of algae can survive and can lead to large algae growths that can create smelly and unsightly mats of vegetation. The ability of the introduced mussels to colonize and reproduce rapidly also creates an economic disturbance to communities that must regularly clean water intakes or develop ways to keep mussels from colonizing intake pipes.

Values and Assets

As this brief history illustrates, human occupancy of floodplains and riverfronts is deeply rooted in historic patterns of rural and urban land use, and in cultural values attached to these environments and uses. As is common in a complex society that has evolved over a long period, some of these values conflict with others. We have exploited the short-term economic values of rivers as conduits for waste, as sites of cheap, easily developed floodplain land or of investment in rapidly appreciating riverfront view property, while tending to discount long-term or externalized costs associated with flood hazards, erosion, pollution, and habitat destruction. A more dispassionate view must recognize values related to legitimate and necessary human uses of floodplains and riverfronts that are sustainable and don't destroy those environments.

Agriculture

Values related to river uses must include the fundamental, original use of floodplains for agricultural purposes. Floodplains provide the most important and extensive areas of prime agricultural soils in most parts of the United States (the major exception is the glacially deposited till plains of the Midwest and eastern Great Plains). The single most productive agricultural region of the United States, the Central Valley of California, is largely a floodplain of the Sacramento and San Joaquin rivers and their tributaries. The Central Valley is also experiencing some of the most rapid rates of conversion to urban use in the country. The somewhat-comparable Willamette Valley in Oregon has, until now, constrained conversion by state-mandated growth management (a program now threatened by passage of a property rights initiative requiring government compensation for regulatory actions).

Agricultural use, especially with conventional, still-prevalent practices, does not mean low impact on physical and biological systems of floodplains and rivers. When forested floodplains are converted to agriculture, rates of surface runoff can increase dramatically along with lowered water tables, soil loss, and sedimentation of streams. Draining wetlands and channelizing streams to drain fields further contribute to heightened floods downstream. Overuse of fertilizers leads to excessive nutrient enrichment in streams. Herbicides and pesticides are sources of pollution. Irrigation, with either surface water or groundwater, reduces streamflows through evaporation or transpiration or through the lowering of the water table. Improper irrigation leads to soil salinization and concentration of pollutants. Mitigation of these many effects is beyond the scope of this book. However, a science of sustainable agriculture that addresses all of these issues is under active development and should be incorporated in any watershed-level planning that encompasses agricultural uses of floodplains.

Commodity Transport, Navigation, Power Production, Fisheries, Water Supply

In many cases the uses of floodplains and riverfronts are closely tied to economic uses of rivers themselves, as illustrated by examples in this section. Continued use of some rivers for barge transportation of bulk commodities was noted earlier, a major example being the transport of Appalachian coal to power plants in the Ohio and Mississippi valleys. This is the major justification for the system of low dams and locks that ensures almost year-round navigation on the Ohio and upper Mississippi rivers. A negative aspect of this is the necessity of locating power plants every 50 miles or so along these rivers, where they can be supplied with coal and can use the river to supply cooling water. The cumulative hard edge of these large plants and their associated ports and coal storage facilities is significant. Air pollution consisting of sulphur and nitrogen oxides from these power plants is transported long distances by prevailing winds, causing acid rain and acidification of lakes in the northeastern United States and Canada. Other negative consequences result from coal strip mining, and more recently mountaintop removal, above the Appalachian headwater streams of the Ohio River.

The channelized lower Missouri River is used for barge shipment of grain. The lower river has been straightened, narrowed, and deepened through bank stabilization and use of groins or dams extending partway into the river. This has cut off connections to back channels and floodplains that help the river transport sediment and nutrients, store high flows, cleanse water, and recharge aquifers. Upstream dams constructed for flood control are managed to ensure flows adequate for continuous navigation. However, barge transport has declined while demand for recreational uses of the river increases. This requires revising the management of water flows to mimic natural conditions conducive to restoring riverine habitat. This modified flow regime is necessary to restore side channels and wetlands in the floodplain. There are currently proposals, and some limited ongoing projects, to make these changes.

Until quite recently, major federal investment has continued to be made in navigation-related modification of rivers, some of it enormous in scale and involving significant effects on floodplains. An example is the Tennessee-Tombigbee Waterway project, which links the Tennessee River and the Gulf of Mexico with 450 miles of canalized or straightened and dredged rivers with dam and lock systems. Designed to provide a navigation alternative to the Mississippi, since its completion in 1985, the Tenn-Tom Waterway has not achieved projected levels of use. Constructing canals and lock systems on the Red River in Louisiana was completed in the 1990s, and the USACE is still putting forth similar proposals for rivers such as the White in Arkansas.

Economic use of rivers for hydroelectric power production and other purposes served by large dams is another major subject beyond the scope of this book. However, hydroelectric power; recreation; wildlife habitat including spawning grounds and passage for anadromous fish; water supply for agricultural, industrial, and domestic uses; and effluent transport all represent potentially competing and conflicting values.

Habitation, Land Uses, and Amenities

Human settlement and habitation in river communities will continue. There is no overriding national or state policy to divert settlement and urban growth away from rivers. Because of the National Flood Insurance Program (NFIP), floodplains themselves have been treated somewhat differently in policy from more elevated riverfront or river valley environments. By limiting some property-owner risk through insurance and thus providing a sense of security that is not always realistic, NFIP has perpetuated the development of river environments.

People want to live in river communities today for the same reasons earlier generations have. They provide economic opportunities and, increasingly, they are valued for the range of amenities they offer. These include views, proximity to nature and wildlife, and various onshore and offshore water-related recreation activities. In addition to environmental, scenic, and recreational amenities, river communities offer a sense of place-based cultural identity reflected in many facets of community life. Festivals and other public events celebrate the community tie to the river. The riverfront may also include physical assets such as buildings, bridges, landings, wharves, riverboats, historic sites and districts, parks and other developed open spaces.

The planning challenge for river communities is to resolve conflicts among these uses. More important, it is to minimize or mitigate the effects of human uses so the natural systems that created these values can be sustained. This usually involves trade-offs. In general, the types of uses that should be maintained and protected are land uses and their built assets that must be near a river, and have either a present economic or public value or a recognized historic or cultural value. Protecting these assets from flood damage may require localized structural solutions. However, environmental degradation and the cost of forgoing potential natural system restoration opportunities must be weighed in every case.

In general, land uses and physical structures that are not river-related or river-dependent should not be located near the floodplain. This is especially true if they negatively affect natural river processes. Even if they do not, these structures occupy riverfront space that many communities would value more highly for river-related purposes and amenities. Contrary to this, however, is the predilection of some communities to allow large-scale commercial developments that are not water-body dependent, such as corporate offices. Some views and amenities may be provided for office workers, but often at the expense of views and access for the larger community and perhaps the sacrifice of historic structures.

Many river communities have shoreline industries that at one time were dependent on the river for water, power, transportation, or effluent transport, but no longer need the river because of technological changes. It may be appropriate to use long-term planning to remove or relocate these structures. Or, they may offer opportunities for adaptive reuse suited to a riverfront area. Converting former industrial buildings or warehouses to mixed-use, retail, entertainment, and tourist-oriented uses that serve to draw people back to the river are a common type of such adaptive reuse that has been successful in a number of cities. An example is the Laclede's Landing historic district on the St. Louis riverfront.

Planning is a value-based enterprise that must weigh, choose among, or attempt to optimize multiple, sometimes competing or conflicting values. Planning in river cities and communities is dependent upon an educated and involved public aware of the full range of values and assets we associate with a river and of the relationships between values that relate both to human uses and to environmental sustainability. Where the twentieth century was the age of big projects like dams and levees, the twenty-first may see the dismantling of many of those projects to use the natural advantages of riverine processes. Science, not public opinion or past practices, should drive future flood management practices.

Case Study: Chicago, Illinois

We can use our six questions to address use-related values and assets as well as natural systems related values and assets. In this case, we will look at the Chicago River in the late nineteenth century.

Chicago took advantage of an emerging cattle market by constructing a stockyard that was centrally located, had good rail transportation, a large labor force, and available water to both feed the cattle and remove the wastes by using the Chicago River to drain waste products away from the site into Lake Michigan.

This led to massive pollution of the river after a few years. These by-products were tainting the city's drinking water, which came from Lake Michigan. The initial value in the river was to take advantage of its ability to flow away from the site. The new value was to restore the importance of Lake Michigan as a clean water source for the city.

1. What Values or Assets Do You Want to Protect or Enhance?

The city needed a clean water source.

Historically, Chicago was a wetland that could process considerable biological wastes. However, much land was drained in the creation of the city and farmlands. The drainage basin is very flat near Chicago and extends just a few miles westward. Biological processes could process wastes but were quickly overwhelmed.

2. What Are the Apparent Risks or Opportunities for Enhancement?

Initially the risk was not having waste removal capability. However, after using the river to solve that problem, the risk was then not having available clean water and the problems that come with contaminated water—disease, loss of economic growth, and so forth.

3. What Is the Range of Risk-Reduction or Opportunity-Enhancement Strategies Available?

The Great Lakes were carved by glaciers and are filled with residual meltwater. The Mississippi drainage basin lies just a few miles away. The United States had extensive experience in building structural projects such as the Erie Canal. The USACE was created to undertake large structural works. These were capabilities available to Chicago.

- The stockyards could be closed.
- A different way to dispose of the effluent could be built.
- Another drinking water source could be developed.

4. How Well Does Each Strategy Reduce the Risk or Enhance the Resource?

Each of the three strategies could result in cleaner water.

5. What Other Risks or Benefits Does Each Strategy Introduce?

- Closing the stockyards would mean an immediate economic depression in the Chicago area and would leave ranchers without a market for their product. It would probably move the pollution problem somewhere else.
- An alternative way of disposing effluent would be a great benefit to Chicagoans and ranchers, but would push the problem somewhere else. It would also be very expensive.
- Finding another drinking water source upstream of the Chicago River would be expensive.

6. Are the Costs Imposed by Each Strategy Too High?

At the time, such decisions were often made based on raw political power.

Downstream users were affected, but were not included in the decision-making process, because they had no political standing. Waste processing technology did not exist.

Future users (us) were not considered. Today the slaughterhouses have long gone, but Lake Michigan may be experiencing a net loss of water. Pollution from the northeastern drainages, including that from the now reversed Chicago River watershed, which drains into the Mississippi watershed, has created a dead zone in the Gulf of Mexico.

7. The Decision

The chosen strategy was to flush wastes from the Lake Michigan basin into the Mississippi and to do it quickly—while there was support within the Illinois legislature and before downstream interests could gain political support.

Within a couple of years after the wastes were diverted, the number of typhoid cases increased in St. Louis. That city sued but lost when the judge found inconclusive evidence. The outcome might be very different today.

Chapter 6

Approaches: Structural and Nonstructural

The most cost-effective approaches to floodplain management are those that incorporate physical, chemical, and biological processes while acknowledging and accommodating human uses of rivers. The whole watershed, including the floodplain, river corridor, rural areas, and urban areas must be managed. Historically, massive structural measures with very large environmental footprints have been the flood management tools of choice, but they present problems. Nonstructural approaches, which work with natural processes, have many advantages. Different approaches can be applied in different parts of the watershed. Appendix B offers a graphic tool kit for floodplain design to help illustrate potential choices.

Managing Change

As a country, we are often enamored with technology and big structural works aimed at "fixing" the river. From the early through mid-twentieth century, flood hazard reduction approaches were dominated by large structural works: major dams on large rivers; many hundreds of smaller dams on most river systems throughout the United States; massive levee systems extending continuously for hundreds of miles along large rivers and their main tributaries; and channelization of many waterways, including entire rivers such as the Chicago River and urban segments of others that, like the Los Angeles River, were transformed into concrete-lined ditches or buried underground.

Sometimes these projects were carried out as much for job creation as for water control. In other cases, single uses such as navigation, irrigation, or power production were the dominant purposes. Even in those

93

projects, planners typically would tally a long list of supposed purposes to produce the most optimistic outcomes in benefit-cost analyses.

Geographer Gilbert White, sometimes called the father of floodplain management, argued for focusing on reducing the effects of flooding rather than changing the river's behavior. The imagery of the terms structural (the Grand Coulee dam) and nonstructural (elevated homes) was helpful in selling the message. Since the last decades of the twentieth century, under the influence of White and his followers, greater emphasis has been placed on nonstructural methods that focus on reducing risk or vulnerability.

The floodplain management community has defined structural measures as those involved in holding back floodwaters or their secondary hazards such as transported sediment. In doing so, structural measures change flood characteristics—the hazard profile. Structural measures include dams, levees, floodwalls, and detention basins. Nonstructural measures seldom significantly alter the river. They include elevating a structure, moving vulnerable items like furnaces above the level of flooding, and using building materials and methods that are water resistant.

Today the concepts of structural and nonstructural are more flexible. Structural approaches include bioengineered structures that look nothing like a concrete dam or levee. Using fill to elevate a home is a nonstructural measure, even though the fill can change the dynamics of a stream. The terms are still used in floodplain management and, in general, structural approaches have a much larger effect on the environment.

There are four ways to approach flood management:

- *Prevent* downstream adverse effects by using existing or enhanced forests, wetlands, and detention ponds. Though the term *prevention* is used by many in the flood management community, it is more accurate to think of using these techniques to *minimize* flood effects.
- *Retreat* from the floodplain and let natural processes work.
- *Accommodate* the effects of floodwaters by elevating structures (without using fill) or using water-resistant materials and building practices.
- *Protect* an asset with measures like strategically placing large woody debris (LWD) to deflect hazardous flows around it, a ring dike around it, or, if absolutely necessary, building a levee, dike, or dam.

We begin this chapter with the authors' acknowledged bias against massive projects that leave big footprints. Often the secondary disturbances and adverse effects they cause over the long term outweigh the project's intended benefits. That said, there are appropriate ways to employ engineering and to combine structural and nonstructural approaches.

Problems with Structural Approaches

Too often we have adopted a quick-flush approach to manage runoff and floodplain flooding problems. These methods use the concept of *conveyance*, or attempting to convey peak flows past each community as quickly as possible. Such efforts inevitably push higher waters onto downstream neighbors. A quick-flush system typically confines the stream with levees and straightens the channel, which quickens the passage of water through the stream channel and increases floodwater depths. Because the faster passage of water creates increased erosion and scour, the channel is often armored to slow down these processes. In addition to passing flood peaks downstream, other problems include loss of water storage, habitat destruction, high maintenance costs, and encouraging floodplain development behind levees that often faces higher long-term flood risk.

Levees, whether made of concrete or bioengineered with a more natural-looking profile, have been the flood management tool of first resort in many places. They confine the river to a narrower channel than its natural floodplain so the watershed loses some of its capacity to store water. The effects of flooding are simply pushed downstream, which creates pressure to add more levees to continue to push the water downstream. The river can no longer migrate across its natural extent, which restricts the development of oxbows, migrating meanders, and other riverine processes that produce the side-channel and off-channel areas that promote a healthy ecological web of life. The complex web of plants and animals simplifies because many of them can no longer live in the levee environment. This not only reduces recreational and aesthetic uses of the river, but it also diminishes the capacity of the river itself to regulate its flows and provide natural flood control.

Continuous levees, floodwalls, or bulkheads; long stretches of hardened channel employing concrete, riprap, gabion walls, or other armoring; and other structures such as groins or wing dams that over time produce a narrowed, stabilized channel are all designed to implement the quick-flush philosophy. When implemented throughout a watershed, they produce higher peak flood elevations in tributary and mainstem rivers.

In worst-case events, this structural system gets overwhelmed and causes catastrophic failures of many of these individual structures. As seen in the Mississippi watershed, communities come to believe levees are absolute protection and continue to build in flood-prone areas. The resulting damage can be worse than if the levees did not exist.

Excessive use of storm sewers and pipes is an example of quick-flush engineering often seen in small, urbanizing sub-basins. They rapidly gather and convey runoff directly to surface streams, contributing to higher flood peaks downstream. Extensive storm sewer systems also cause increased velocities, erosion, channel instability, and pollution in the streams they feed. The magnitude of these effects is often overlooked. Theoretically, if all the surface area of a sub-basin is served by a storm sewer system, the peak flow from the sub-basin can increase as much as 100 percent.

Water spreading is a fundamentally different philosophy than conveyance. A water spreading system puts water into temporary storage to reduce downstream flood peaks. Holding water back, storing it on the land, and slowing or lengthening its flow path are among the approaches used. This process allows water to infiltrate valley alluvium and potentially to recharge underlying aquifers. The slow release of this water improves base flow and maintains water tables, which can help ensure sustainability and continuity of local water supplies and support riparian vegetation. Healthy riparian vegetation in turn provides further friction to slow the water and increase storage, a positive feedback loop in a water spreading system.

As flood-control devices, dams impound and store water but they also serve to maintain water levels and water availability for navigation, irrigation, water supply, hydropower, and other purposes. Against these various potential uses (and often only a single use dominates in any one project) are an array of environmental impacts, costs, and externalities. Flooding impounded river valleys causes loss of riverine, riparian, and floodplain habitat, and loss of land highly suitable for agriculture and other purposes. Anadromous fish are blocked from their spawning grounds.

As serious as these upstream effects are, in recent years we have come to realize that the downstream effects of dams can be equally profound. Trapping sediment behind dams starves sandbars and beaches, causing their disappearance. Channel degradation also results, leaving enlarged alluvial fans where tributaries enter, blocking or confining flow in the river's channel. Vegetation that relies on flooding for reproduction, such as cottonwood trees, begins to die out. In addition, release of colder, less turbid water from behind dams has widespread effects on fish and other wildlife.

Many smaller dams no longer serve their original purposes, or provide benefits that are minimal in relation to effects, and are being removed. Whether any more large dams should be built for flood management might be a moot question, because relatively few suitable, unexploited sites remain in the lower forty-eight states and present-day economics do not favor such large projects. With changes predicted by climate models, however, there may again be advocates for such projects.

This is likely to happen in regions where climate change clearly exacerbates floods or threatens water supplies. In such cases, dam construction may be seen by some as a panacea. An example of this is the Cascadia urban corridor extending from Eugene, Oregon, to Vancouver, British Columbia. A majority of the nearly 10 million people now living in this corridor are dependent on water supply from relatively small impoundments on the west slope of the Cascade range. Much of the stored water is provided by mountain snowpack that gradually refills the reservoirs as melting occurs throughout the spring and summer. With climate models forecasting a greatly reduced snowpack over the next fifty years and a doubling of the region's population predicted over the same time, existing storage will not be sufficient. There will likely be pressure to raise dam heights and build new dams to provide larger reservoirs extending farther up valleys.

In the same region, some communities located on rivers or in floodplains may face more frequent severe floods because of the shift from a snow-dominated to a rain-dominated climatic regime. Damaging floods on Puget Sound–area rivers in November 2006 and again in January 2009 may be an indication of this change. Fall and winter storms deposited rain at even the highest elevations. The lack of snow contributed to immediate runoff. Major floods resulted. This threat, too, may fuel arguments for more and higher dams.

Rethinking Large Structural Approaches

In certain situations, at certain scales, given certain existing conditions, and when used in combination with nonstructural measures, there is a place for structural approaches. They can be useful and their negative effects can be minimized.

One example is where development or essential infrastructure is threatened by erosion or channel migration. Structures can be designed for limited sections of channel to resist lateral movement. Even in this situation, however, a total hardening of the riverbank is not always required, and a better solution may be found with engineered placement of woody debris to provide friction as well as stabilization.

Another example is using limited sections of dikes or levees to provide an intermediate level of flood protection for highly vulnerable development or infrastructure. The key word is limited—in both scale and linear extent. This might be an appropriate place for a ring dike. It would surround the immediate area at risk, as opposed to a continuous levee along a substantial stretch of river.

Over time, existing levees can be reconfigured to transition from a quick-flush (conveyance) to a storage (water spreading) approach. Moving levees back from the river wherever space is available is a fundamental step, as shown in Figure 6.1. Traditionally, setback levees have been used to increase conveyance by producing a larger channel cross section. The real value of this approach, however, lies in the potential for re-creating some natural floodplain within the space created by the levee setback. Ideally, this space should be sufficient for the river to reestablish a degree of meandering with side channels, wetlands connected to the river, and vegetated channel edge throughout. Habitat for native plants and animals can also be reintroduced and provide some level of flood protection, as well as encourage increased numbers of endangered spe-

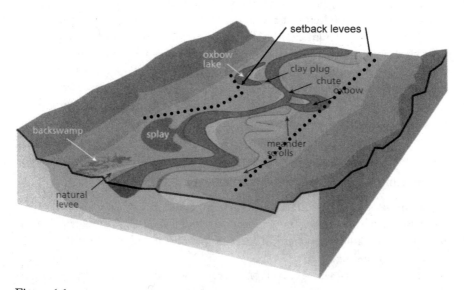

Figure 6.1
If levees are required, they should not follow the individual meanders, but should be setback levees, as shown above, to allow the river to continue its natural meandering. (Source: Adapted from the NRCS *Stream Corridor Restoration*)

cies like salmon. To some degree, these features can be engineered at the outset, being cognizant of the need to let natural processes take over to re-create the riverine landscape.

Another approach to retrofitting existing levees is to lower them at strategic locations, designing them to be overtopped by floodwaters. This can occur where flood-tolerant vacant land exists behind the levee, such as little-used farmland. Prior agreement can be reached with land-owners to periodically use this land for storage. Overtop levees can be designed to release water gradually, without erosion, sedimentation, or other damage to either the levee or the adjacent land used for storage.

Dams can also have a limited role among these revised approaches to flood hazard management. Small check dams strategically located throughout an upper watershed can create storage through a system of scattered impoundments or diversion of water into storage areas. Where existing larger dams are kept, their flow regime can be managed to more closely mimic natural flows, such as seasonal flood pulses. This still pro-vides some flood protection while reducing negative downstream effects.

A special system of dry dams can help communities in a watershed of generally low topographic relief that face a major flood threat from a smaller river. Such a system has been implemented in a few places, the prototype being the Miami Valley Conservation District upstream of Dayton, Ohio. These dams do not impound water under normal condi-tions, and so they have a minimal effect on natural systems. Their only purpose is to impound water to mitigate peak flood hazards.

Native Americans warned early settlers that the floodplain site of Dayton was a dangerous place to build. Although the Great Miami River is not large, it has a sizeable upstream watershed and three tribu-taries join within 1 river mile above the site of downtown Dayton. Floods were frequent throughout the nineteenth century as most of the watershed's forests were cleared for agriculture. March 1913 saw the greatest floods of record throughout the Ohio Valley. Most of Dayton was inundated, the downtown flooded to the third story of buildings, and 700 people died.

Early-twentieth-century Dayton was a leader in innovative technol-ogy and the response to this event was remarkable for its time. Civil en-gineer Arthur Morgan was hired to find solutions. His first action was to draft state legislation, the Ohio Conservancy Act, that allowed local governments to establish conservancy districts with taxation powers to address flood management. The Miami Conservancy District, first of its kind in the United States, was established in 1915, with Morgan as its first director. He advanced a technological solution, based on a system of what he termed dry dams on the Great Miami and several tributaries.

These dams have slots that permit normal passage of water and fish under ordinary conditions without impoundment or channel modification. Areas behind the dams remain in agricultural use or forest. During peak flood events these dams close and water is impounded temporarily behind the dams or diverted onto neighboring fields. For eighty years the system has functioned successfully, and Dayton has not experienced significant flooding.

The dry dam approach has been implemented in a few other places. One example is a U.S. Army Corps of Engineers project on the Chena River upstream of Fairbanks, Alaska, built in response to a major Fairbanks flood in 1967. This system is similar in many respects to the Dayton example, but also incorporates a provision for major diversion of floodwaters into an adjacent, larger water body, the nearby Tanana River. In this respect, it is also similar to diversion structures that have been built in other places. The most famous example is the Bonnet Carré spillway on the Mississippi above New Orleans, constructed in the 1930s to divert floodwaters into Lake Pontchartrain before reaching the city. This structure has been used only twice for that purpose, both times successfully.

Yet another area where technology may be able to play a larger role is in underground storage of runoff through facilitated recharge of aquifers. Systematic management of aquifers and the technology for artificially recharging groundwater reservoirs is still at an early stage. In the Los Angeles area, recharge basins have been constructed to capture peak winter flows from the mostly channelized rivers that cross the urbanized plain between the mountains and coast. This capture avoids some urban flooding, helps to offset declines in the water table, and prevents saltwater intrusion.

Clearly, such recharge systems must be designed for the specific hydrologic, topographic, and soil conditions of a given area. Underground storage in the upstream portions of a watershed that has high relief may require targeting very specific pieces of terrain, such as gravel deposits at the base of mountains where stream gradient flattens. In the Puget Sound area, for example, glacially deposited gravels at the base of the Cascade range offer potential access to large aquifers thought to exist in the area. However, information on the extent of these aquifers and the optimum locations for recharge is currently insufficient.

California's Central Valley

The Mississippi is not the only watershed that relies extensively on levees. The Central Valley of California faces significant flood risks. In January 1997, more than thirty levees ruptured and 300 square miles in the Central Valley were inundated by floodwaters. Forty-eight counties were declared disaster areas. Nine people died and 128,000 people were evacuated. Damage approached $2 billion and estimated indirect costs exceeded $5 billion.

A 2007 report to the California Department of Water Resources by an independent review panel explored the twin problems of increased flood risk due to development and predicted climate change, and aging levee systems that may fail.

From the report:

> Recent inspections have raised serious questions as to the integrity of many levees that protect communities and property in the Central Valley. Conservative estimates of potential direct flood damages in the Sacramento area alone exceed $25 billion. In some areas of the Central Valley, communities would experience flood depths of 20 feet or more when the levees fail. . . .
>
> Many of the system's levees were poorly built or placed on top of inadequate foundations; others have been inadequately or intermittently maintained. In addition, efforts to protect the Central Valley from flooding have also significantly degraded the natural and beneficial functions of the rivers and their floodplains, threatening the loss of species, destroying habitat, and failing to take advantage of the floodplain's natural capacity to store floodwaters and to recharge aquifers below them. The current flood-control system of the Central Valley is incapable of dealing with the threat of severe flood events, placing its urban centers at considerable risk while incurring significant environmental costs.

To fix the system, the report recommends both structural and nonstructural approaches. Whenever possible, levees should be set back from the river; development on floodplains should be discouraged; floodproofing, elevating homes without using fill, and other nonstructural approaches should be encouraged.

Nonstructural Approaches: Low-Impact Development

Some low-impact approaches can reduce the effect on, and the consequences of, development in watersheds and floodplains. These approaches vary depending on the scale and geographic area within the

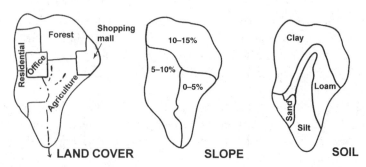

Figure 6.2
Many watersheds encompass a variety of land cover, slopes, and soils types. Adverse effects from development can be minimized if the watershed is looked at as a whole. Flood-reduction and water-enhancement measures can be taken throughout the watershed.

watershed at which they are applied. Figure 6.2 shows some of the watershed factors to be taken into account, including existing land cover, soil, and slope. Low-impact development sustains natural system functions and reduces risk in both higher elevation portions of river basins, watersheds and sub-basins, and in downstream floodplains and river corridors. Reduced-impact development and urban design reduce risk and enhance use while bolstering historic, cultural, and amenity values in urban riverfronts.

Higher Elevation Portions of River Basins, Watersheds, Sub-Basins

Much can be gained by focusing on the upper watershed, including smaller elevated sub-basins adjacent to floodplains, using them like a kind of water tower for downstream communities. It is possible to maintain upstream storage and infiltration through natural systems such as wetlands and forest cover. This can be done by managing land use and land cover and employing development practices that limit the percentage of surface runoff. These measures can limit increases in peak flows while concurrently reducing dependence on upstream dams to reduce flood hazards.

Downstream Floodplains and River Corridors

In many places, urban development dominates downstream floodplains and river corridors. Urban land cover affects a number of factors, includ-

Table 6.1
Impervious Surfaces Change Predicted Rates of Runoff

% Impervious surfaces	% Surface runoff	% To groundwater	% Absorbed	% Evaporated
0	10	25	25	40
10–20	20	20	20	40
35–50	30	15	20	35
75–100	55	5	10	30

ing the extent of impervious surfaces and their spatial distribution in relation to slopes and soil types. Table 6.1 shows the importance of impervious surfaces in disturbing natural drainage patterns. In this simplified example, a fully urbanized city causes 55 percent of precipitation to become surface runoff, greatly increasing the frequency and size of floods. In addition, groundwater receives very little recharge, depleting aquifers.

Key principles include:

- Maintaining post-development runoff at predevelopment levels.
- Minimizing the impervious footprint of buildings, parking lots, and streets.
- Allowing for storage and ground percolation.
- Slowing and extending the path taken by surface runoff.
- Maximizing the use of trees and forests everywhere possible.

Drainage paths can take the form of a preserved or restored natural drainage network on the surface, or a constructed system of swales. Benefits can be enhanced through circuitous routing, increasing detention and percolation along the paths, and maximizing the roughness of channel surfaces. The net effect should be no increase in flooding, and, where possible, an increase in storage.

Onsite detention and/or retention through preservation or restoration of natural wetlands, or through engineered systems employing wet well or dry well technologies, can also minimize the effects of development. Buried drainage in pipes, culverts, or storm sewers generally has the opposite effect and can contribute to substantially increased flood heights downstream.

Figure 6.3 illustrates a general concept for low-impact suburban residential development within a small sub-basin. Each home is shown with a tree to emphasize the importance of using natural biology to help dampen high water flows. Development is designed to minimize the number of impervious surfaces and to mitigate their effects. Overland

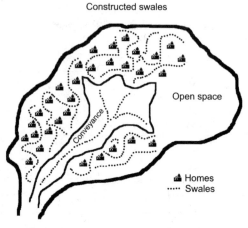

Figure 6.3
Careful planning can ameliorate the effects of suburban development. In this small basin, an engineered swale system maximizes the runoff path of water, and dedicated open space along the conveyance channel allows for water storage.

drainage flow from developed lots and impervious surfaces is routed through a system of swales. They lead to a central conveyance zone that follows the natural stream channel and floodplain, exiting the sub-basin into the next larger watershed.

A swale system should use natural drainage paths where possible. However, it can also include constructed swales that can serve other open space uses under normal runoff conditions—greenbelts, walkways, bike paths, and wildlife corridors. Swales can be designed with underlying soil, sand, and gravel layers to maximize percolation, and with surface vegetation that maximizes roughness and minimizes flow velocity and erosion. The central conveyance zone is likewise kept in open space with protected forest, wetland, or other naturally occurring vegetation and its consequent woody debris. All these elements work together to minimize overland runoff, increase percolation and groundwater recharge, minimize flood peaks, and protect wetlands, riparian zones, and hyporheic zones.

Open Space and Storage Opportunities

In all cases, land-use planning for floodplains should emphasize flood-tolerant open space land uses, especially those that preserve natural vegetative cover such as nature preserves, wildlife refuges, wetland areas, and parks and recreation areas oriented to passive, nature-directed activities. Agriculture, though it has significant impacts on hydrology and aquatic biosystems, represents the dominant, established land use in many flood-plains. Converting from agriculture to urban uses constitutes a major disturbance that can destabilize existing natural systems. Also, agriculture is more tolerant of flood conditions than urban areas. These are practical reasons for conserving and maintaining existing agriculture in floodplains where restoration to natural conditions is not feasible. Agricultural zoning has been an effective tool for accomplishing this in many places. Other land-use management tools, such as development rights and conservation easements, can also be useful.

Other Land-Use Approaches

Other concepts for land-use management in floodplains include:

- Conserve or restore *natural storage* within the watershed, floodplains, and channel. Wetlands protection through critical areas or similar local ordinances is a key element. Also important are aspects of stream restoration that reconnect a stream to side channels, abandoned channels, oxbows, and other portions of its floodplain.
- Create *buffers* where structures are required to be set back from the bank of a stream or river. In part, this coincides with the FEMA floodway development restrictions (see Appendix A). However, buffers should be designed to serve many purposes in addition to allowing conveyance. Buffers can protect native riparian vegetation that can filter pollutants from runoff into the stream, provide nutrient sources for the stream, and modulate stream temperature by providing shade. Buffers can also protect a significant portion of a stream's hyporheic zone, allowing water to pass between that zone and the surface. The width of required buffers should be based on local physical and environmental conditions.
- Establish *habitat corridors* to allow a riverine community to develop and provide natural flood management. These corridors can also help protect endangered species.
- Establish *floodplain zones for flood frequencies less than 100 years* (e.g., ten, twenty-five, or fifty years). Use different land-use controls or development mitigation requirements for each zone based on the differing levels of hazard. For example, a retreat policy could be

government buyout and removal of structures in the one- to ten-year zone. Accommodating floods could be done by elevating existing development and prohibiting new development in the ten- to twenty-five-year zone, or regulation based on a density fringe floodways concept in the twenty-five to fifty and fifty- to one-hundred-year zones (for an explanation of density floodplains, see Appendix A). This would require periodic updates of floodplain maps because changing watershed conditions can change floodplain boundaries. The production of digital maps through FEMA's Map Modernization program should allow for quicker revisions.

- *Discourage use of fill.* Encourage the elevation of buildings using open-story ground floors that do not obstruct floodwaters. Remember, elevated structures in an earthquake zone must be strong enough to resist seismic forces.

- *Identify depth and velocity zones.* Regulate development in these zones. The FEMA regulatory floodway is not necessarily the only area where swift floodwaters pose a danger to life and property. Other overflow channels can carry deep or high velocity flows. Building elevation requirements may not be sufficient to provide protection, and development prohibitions or special engineering requirements may be needed.

- Use *setback levees or dikes.* These increase conveyance areas and allow a river to reestablish more natural conditions in the near-channel portion of its floodplain. They allow the channel to migrate and reconnect to back channels and wetlands. A broader floodplain will distribute energy over a wider floodplain and reduce erosion. Also, levees and dikes can be lowered or redesigned to overtop at certain flood-tolerant locations during peak flows of major floods. This will reintroduce flood pulses to adjacent land and lower the downstream flood peaks. Coordinating levee and dike systems throughout a floodplain can equitably distribute flood effects.

- Develop *basin-wide, computer-based hydrologic models.* These models can be employed as regional land-use planning tools. Using them for analysis makes it possible to tailor planning to projected physical characteristics, land use, and development of each basin. It is important, however, to remember that models must be used in conjunction with field studies to accurately understand a basin.

- Use *regional greenways and greenway systems.* These can promote citizen interest in buffers, setback levees, wetlands, and riparian and other open space protection. They preserve corridors of open space and can provide a variety of recreational uses, allow conveyance of floodwaters, protect natural systems, and preserve scenic, cultural, and historic resources along a river. Greenways fit into both rural and urban environments, providing connection between the two.

- *Limit hard edges.* If banks must be armored or levees built, restrict structures to selected reaches or to only one side of the river.

Urban Design Approaches for Urban Riverfronts

Urban design is a branch of urban planning that deals with the spatial, functional, visual, aesthetic, and psychological interrelationships of built form, open spaces, and natural environment. Urban design offers many tools that can be applied to the problem of effectively integrating an urban environment with a river environment.

The following are urban design and development approaches, many of which have been used in U.S. cities:

- Remove or avoid building levees or floodwalls in the central portion of the urban riverfront to preserve the city's window on the river. Example: Davenport, Iowa.
- Where flood protection is required in the central riverfront, replace the floodwall or existing levee with a gradually sloping levee surface and winding pedestrian pathways, overlooks, and green space oriented to passive recreation. In this manner, it is possible to employ a levee that does not look like a levee and still invites access to the river. Example: Gateway Arch Park, St. Louis, Missouri.
- Where existing levees cannot be removed, develop pedestrian access across and along the top of the levee with walkways, viewing platforms, directional signs, and interpretive displays. Examples: the Moonwalk in New Orleans, Louisiana; Savannah River levee in Augusta, Georgia.
- Put highways paralleling the river below grade, lid portions of them, and develop lids with uses appropriate to a riverfront district. Example: Cincinnati, Ohio.
- Move or site parallel highways back from the river, preserving riverfront districts. Example: Old Sacramento, Sacramento, California.
- Remove riverfront highways and replace them with parks and open space. Example: Tom McCall Riverfront Park, Portland, Oregon.
- Preserve and restore historic riverfront districts or representative portions. Find adaptive uses for buildings that are economically viable. Preserve historic riverfront artifacts such as early bridges and docks.

Provide localized structural flood protection as needed. Example: Laclede's Landing and the Eads Bridge, St. Louis, Missouri.

- Connect and incorporate urban riverfronts into citywide, metropolitan, and regional greenway systems. Example: Willamette River
Greenway, Portland to Eugene, Oregon.
- Create a regional River Edge Authority with broad powers to acquire,
preserve, restore, and develop for public purposes lands bordering both
sides of a major river along its course through a metropolitan region.
Objectives include reducing risk through wetland floodwater storage
and naturally vegetated river edges and buffers, habitat protection and
restoration, recreation, visual amenities, and public education. Example: Meewasin Valley Authority, Saskatoon, Saskatchewan, Canada.
- Create and sustain citizen interest in the riverfront as a defining sym-
bol of community place. Example: San Antonio, Texas, River Walk.
- Connect the channel overflow paths laterally to the floodplain. The
more reaches that are currently unconnected, the more significant this
can be.
- Move levees or floodwalls back from the river and reserve the riverfront area for open space and other flood-tolerant uses. This approach
is especially suited to restoring a soft edge.
- Control building heights and the orientation of buildings to preserve
views of the river from the city, and of the city from the river.
- Limit development to river-related uses. Clearly, this requires a broad
definition and can include everything from port activities, marinas,
water-based recreation, and historic districts to uses that reflect historic and cultural values related to the river.
- Use riverbanks and shorelines primarily for passive recreation activities that do not require paved surfaces, structures, or hard edges. Expand linear soft edges to the maximum extent possible within the urban area.

Graphic Tool Kit

When analyzing current conditions or planning for future changes,
graphic notation is extremely useful. Using either geographic information
system (GIS), computer, or hand drawings, watersheds can be delineated
into areas of retention, detention, discharge, floodplains, and active terraces. These maps can then be manipulated to show potential changes in
the built environment, biological conditions, or other physical conditions.

As various strategies are discussed, maps can be updated to show changes in the watershed. This can be used for flood-reduction activities and water-enhancement projects. Appendix B shows an extensive graphic vocabulary that can be used, along with an example of how it can be applied.

Summary

Our emphasis must be on preserving or enhancing natural risk-reducing processes. When our urban or rural development degrades them, we must replace them—often at exorbitant costs. At some threshold, natural systems collapse and we enter a new regime. At that point, it is virtually impossible to revive natural conditions and a new equilibrium must be established. This nearly always means more expensive treatments for water quality and quantity. In the end, it is usually much cheaper to preserve or enhance natural systems.

Case Study: Buck Hollow River, Oregon

The Buck Hollow watershed is a very erodible landscape where physical impoundments and friction from vegetation to slow the water are vital. Streamflow in the Buck Hollow River was historically perennial on the main stem, but intermittent in the upper reaches. Most of the principal tributaries are currently intermittent. The valleys of the main stem and major tributaries are relatively narrow and confined by steep and high canyon walls with slopes typically greater than 60 percent. The uplands are rolling hills, sharply dissected with deeply entrenched drainage systems. Of the 127,000 acres, 60 percent is in rangeland, 36 percent is farmed, and 4 percent is road and urban uses.

Grazing over 130 years led to the gradual deterioration of the watershed. Summer flows disappeared in the mainstem and lower side drainages. High winter flows with heavy concentrations of agricultural by-products were destroying the riparian areas, exposing excessive sediment, scouring channels, and increasing turbidity.

1. What Values or Assets Do You Want to Protect or Enhance?

Ranchers and other users needed to reduce winter streamflows and increase summer ones. They also needed to remedy water quality problems specifically related to salmon.

2. What Are the Apparent Risks or Opportunities for Enhancement?

The watershed experienced reduced levels of summer soil moisture and unsustainable levels of erosion. Resident and migratory fish had all but disappeared.

There were three main risks to the watershed: salmon would be lost, farm productivity would diminish, and the aesthetic qualities of the watershed would be lost.

3. What Is the Range of Risk-Reduction or Opportunity-Enhancement Strategies Available?

- One option was to do nothing.
- The traditional channel-based approach would include dams on the mainstem and side drains, armoring channel branches, and gravel mining.
- More comprehensive watershed-based approaches would include the use of residue tillage practices, the creation of grassed waterways, sediment basins and terraces, filter strips, critical area planning, and range seeding.

4. How Well Does Each Strategy Reduce the Risk or Enhance the Resource?

- Doing nothing would not reduce the risk.
- Dams, armoring, and gravel mining would tend to keep more water in the river and reduce erosion. Dams would not help with salmon runs as they are difficult for fish to negotiate.
- The more comprehensive watershed-based approaches would address the needs for streamflow, reduced erosion, and increased fish runs.

5. What Other Risks or Benefits Does Each Strategy Introduce?

- Doing nothing would allow the problems to get worse. That could lead to parts of the watershed no longer being usable and making an eventual reclamation project more costly.
- Dams could make the fish runs worse. Structural solutions tend to be expensive and have costly long-term maintenance. Gravel mining

might offset some of those costs. There would be a significant problem if all retention structures within a single drainage failed.

- The more comprehensive watershed-based approaches would also be costly, but are likely to have a longer life span than the structural strategies. Because they work with the natural system of the watershed, they are likely to have a lower total cost over the life of the project.

6. Are the Costs Imposed by Each Strategy Too High?

- Doing nothing would leave individual ranchers and residents to bear the cost of an increasingly dysfunctional watershed.
- Any major change to the river would call for major financial resources, beyond that of any individual or small community. Potential funding sources included grants from NRCS P.L. 566 Small Watershed Grant program and the Oregon Governor's Watershed Enhancement Board (GWEB). Because there was a direct economic value to them, it would be appropriate for ranchers and farmers to bear some of the economic burden. Supporting programs included the NRCS Conservation Reserve Program (CRP) and NRCS Conservation Reserve Enhancement Program (CREP).

7. The Decision

After discounting more traditional approaches such as building dams on the main stem and side drains, a comprehensive, watershed-based approach was decided on. Initially there was considerable skepticism by the residents. Because the strategy required participation by landowners, they had to see a direct benefit for their investments. The agreed-upon approach relied on partnership between the federal government (NRCS), the state (GWEB), and individual ranchers.

The key element of this approach was using a suite of management practices in the uplands to improve the hydrologic conditions of the croplands. These measures included using residue tillage practices as well as the creation of grassed waterways, more than 200 sediment basins and terraces, filter strips, critical area planning, and rangeland seeding.

Since the recovery project began in 1991, wildlife has increased in the watershed because of the expanded areas of upland habitat. Between 1989 and 1995 the rangeland productivity doubled on 20,000 acres. At the beginning of the project there were only a few stands of cottonwoods with no new seedlings, and juniper invasion was severe in many areas. Now the riparian corridor includes alders and willows. Although fish populations are still below potential, steelhead redds are increasing.

The total cost of the project from 1991 through 2005 was $3 million. It appears the watershed is stabilizing. A watershed council was established and will continue to oversee the conservation work.

Chapter 7
Capabilities and Tools

Capabilities are the money, power, and timing of resources available for any solution. Different stakeholders have different capabilities, which must be taken into account. All stakeholders (homeowners, renters, businesses, government agencies, recreational groups, and so on) must be involved in creating and carrying out solutions. Tools are methods of bringing about change. A suite of private- and public-policy tools exists for managing floodwaters and preventing adverse effects. However, they were mostly developed for other purposes and we need to apply them differently. Public policy, laws, and potential funding sources must be directed at taking advantage of natural and beneficial systems.

Capabilities

The resources at your disposal—your capabilities—depend on many things. A government agency that can zone land and buy or build structures has more capabilities than does an individual. However, capabilities can be enhanced by bringing together more people and a wider variety of stakeholders on an issue.

In this book we define capabilities as money, political power, and timing. These three elements will be discussed as part of the tools, approaches, and strategies in the following chapters.

Money

All projects need funding for construction, operation, maintenance, and the costs of mitigating adverse effects over the life of the project, many of which are unknown at the time of construction.

For example, suppose a community engages the U.S. Army Corps of Engineers to build a levee. The USACE requires a funding match for construction and a maintenance agreement from the community. Not covered, but often equally important, the community must accept the responsibility of mitigating effects over the long term. These can include removing debris deposits downstream, riverbed deepening, and losses in resident species.

Nonstructural projects have the same funding needs. However, because they typically address the consequences of flooding instead of changing the river, they tend to have fewer negative effects and have lower maintenance costs.

There are times when funding simply does not exist. In this economic downturn, for example, many government agencies are likely to have their budgets cut—some severely. However, you should continue to act. As Louis Pasteur said, "Chance favors the prepared mind." In this context, it means use this time to explore your options. If you have a well-thought-out plan, especially one with partnerships and political endorsements, you may be at the front of the line when funding opportunities appear.

Political Power

Much of your political power depends on who you are or who you represent. If you are an individual landowner, your capabilities may be limited to addressing flood consequences on the property you own. However, even with this limitation, partnerships can be formed with up- and downstream owners or other interested people. If you create an association of landowners, your power will be greater. Groups of local governments or states can leverage even greater political power.

Timing

Different approaches are available at different times. For example, as previously noted, money may not be immediately available. The cycle of emergency management offers a framework for understanding the timing of some capabilities. Each phase in the cycle of emergency management (preparedness, response, recovery, and mitigation) offers an opportunity to take risk-reduction measures or to enhance our resources. As the effects of climate change become more apparent, more funds should become available and this money could represent a new opportunity.

PREPAREDNESS

Plans and preparations must be made for a disaster *before* it occurs. Typical actions include setting up a system to coordinate emergency management personnel from different jurisdictions, establishing flood forecasting and warning systems, and identifying facilities to act as storm shelters.

RESPONSE

Response occurs *during* the flood. This involves acting to save lives and protect property through warning, evacuation, emergency public information, search and rescue, and health and medical care. As seen in Chapter 1, response activities receive the most funding, even though money spent on mitigation is far less expensive in the long term.

RECOVERY

Recovery actions begin *after* the flood. They are geared toward returning all systems to normal. Short-term activities could include an emergency repair to the water system or crisis counseling for flood victims. Long-term activities work to stabilize systems and include such activities as providing redevelopment loans, legal assistance, and community planning. These measures can be in place for years after a disaster.

The environment must also recover. After a large flood, the river channel itself may change as it destroys old cutbanks, gravel beds, and meanders while creating new ones. Wildlife habitat, such as salmon beds, may be destroyed or created. Flood management systems like levees may need repair. Overtopped levees can be permanently breached or set back to allow the river to develop a more natural meander. This could be an opportune time to undertake restoration activities to create a more natural riverine environment that can lessen the effects of the next large flood.

Recovery does not mean going back to exactly what existed before the disaster. After a disaster like a large flood, both the community and the river recover to a new normal. There is often a desire to return to the comfort of the past, but it may be too expensive or the environment may have reached a new threshold and cannot be returned to that which existed before. The river may have carved a new channel, for example, perhaps reclaiming part of its floodplain.

Many times humans perceive a change as a loss. However, change is also an opportunity to implement new programs that work more closely with the river's natural tendencies. Old practices can be updated. If river communities discuss and plan for their watershed before major floods, it will be easier to make these types of decisions and find funding for them.

MITIGATION

Mitigation activities may begin *long before* a flood occurs or be integrated into the recovery process. Mitigation involves activities that prevent a disaster, reduce the chance of it happening, or reduce its damaging effects over the long term. Typical mitigation activities can include structural or nonstructural measures that make the built environment more resilient. Examples include rezoning, removing homes from floodplains, and elevating structures above floodplains. The recovery phase also often offers extensive opportunities for mitigation. If, for example, a bridge has been destroyed, the recovery phase offers an excellent opportunity to rebuild it stronger.

Recently, the emergency management community has added prevention as a fifth phase. Prevention is the attempt to decrease the likelihood of the hazard occurring in the first place. Prevention measures reduce risk in the location, severity, frequency, or timing of the flood. Prevention measures are often structural and could include levees, dams, and detention ponds. Within the context of prevention, mitigation takes on a slightly different and narrower meaning. That is, mitigation attempts to decrease the consequences from flooding, not prevent the flood itself.

MITIGATION, RISK REDUCTION, AND DISASTERS

Often the opportunities to reduce risk come and go quickly, and many times these opportunities are driven by other agendas and objectives.

Here are some common examples of opportunities for risk reduction.

- A bridge is replaced to accommodate higher traffic volumes.
  Opportunity: Lengthen the spans to accommodate higher river flows or provide more storage, whichever has the more beneficial effect.
- Conversely, a flood may wash away a bridge.
  Opportunity: Build a larger bridge that could accommodate expected higher traffic volumes.

- A flood destroys a home.
  Opportunity: Elevate the structure when rebuilding or relocate it to a higher, less flood-prone location.

Mitigation is often referred to as the cornerstone of all risk-reduction strategies: What you cannot mitigate, you must prepare for; what you haven't prepared for, you must respond to; and ultimately, what you have not responded to, you must recover from.

Note that emergency managers and ecologists use the term mitigation slightly differently. For an emergency manager, mitigation is an action that eliminates or greatly reduces risk over the long term. To emergency managers, moving a home off the floodplain would be mitigation. Taking actions to reduce the effect of floodwaters by moving valuables upstairs would be preparedness.

Ecologists refer to mitigation as a corrective action taken to lessen the effect of risk. Appropriate mitigation for a channel-straightening project might be placing large woody debris to compensate for losing fish resting and hiding areas. To emergency managers, mitigation would be allowing the stream to lengthen, storing water and reducing flood peaks downstream and eliminating the need for a corrective action.

Another way of looking at timing is from the perspective of a building's life cycle. It is sometimes said that the river is dynamic and our built environment is not. This is not entirely true. Buildings and infrastructure decay and need to be repaired or replaced. Lots are assembled and redeveloped. Structures—even whole neighborhoods—move.

As we have discussed throughout this book, change has elements of risk and opportunity. As buildings are repaired, rebuilt, or relocated, there may be opportunities that are only cost-effective at the time of change. Here are some examples.

- Owners want to replace heating system.
  Opportunity: Owners could elevate system elements above flood level.
- Owner needs additional space.
  Opportunity: Owner could elevate the building, creating storage for movable items in the lower, flood-prone area.
- Developer buys megastore along a river and determines the property's best use is actually high-density dwelling.
  Opportunity: Remove fill and build dwellings using low-impact development principles and exploiting riverfront amenities.
- Riverside armoring is degrading.

Opportunity: Create a soft edge, and take advantage of more eco-friendly restoration activities to provide flood storage and reduce downstream flooding.

Tools

There are many private- and public-policy tools available for floodplain management. Here we'll look at tools associated with river ownership, property rights, public policies and laws, and funding.

A tool that cuts across all these lines is planning. Developing a plan through a public process where strategies are widely discussed can increase the number of people and groups that support it and will help leverage opportunities as they become available. Timing was listed previously as a capability; having a plan will help you maximize it.

River Ownership

The question of who owns a river is a complex one and illuminates many of the problems that exist in managing floodplains. It is not the same as the question of who has jurisdiction over a river. Both river ownership and jurisdiction are a complex patchwork of private entities and all levels of government.

Public land surveys of the nineteenth century laid out the Township and Range grid over the United States from the Midwest to the Pacific. At that time, surveyors drew meander lines on each side of a river, generally along the tops of the banks. Behind those lines, most of the gridded land passed into private hands. In most cases, though, the riverbeds were given to the states under state constitutions or charters or were later claimed by the states. States assert regulatory authority over riverbeds and portions of riverbanks, requiring hydraulic permits for such activities as excavation or fill within these areas.

If a stream falls within the definition of a navigable river, then the USACE has jurisdiction over navigation on that river and any modification of its bed or banks including development in or modification of jurisdictional wetlands. A USACE 404 permit is required for changes. These have some hydrologic connection to navigable rivers, although the exact definition of these wetlands has changed several times, subject to political influence. Some states also regulate wetlands. This is just one place of overlapping federal and state authority for managing components of a river's environment.

State and local government control some aspects of rivers. Land adjacent to the riverbank (for example, riparian zones) is likely to be in private ownership but may be subject to local zoning laws. Some states have mandated additional regulation under shoreline or coastal zone management programs or critical areas legislation. The state of Washington, for example, requires localities to develop master programs and special land-use controls for a 200-foot-wide shoreline zone landward from the mean high water line on each side of rivers and streams with 20 cubic feet per second or greater mean annual flow.

Many other public and private entities have ownership or jurisdiction over aspects of a river. The U.S. Bureau of Reclamation owns and operates dams and irrigation projects. Public and private water suppliers and energy utilities own and operate dams, the latter under licensing by the Federal Energy Regulatory Commission.

The U.S. Environmental Protection Agency (EPA) and the states, through discharge permits for point sources of pollution, enforce water quality standards under the Clean Water Act and its amendments. Surface water runoff is managed by local government stormwater utilities.

Biological resources of rivers are managed by the U.S. Fish and Wildlife Service (USFWS) and by state Fish and Wildlife Agencies. USFWS is responsible for implementing the Endangered Species Act.

Public lands along rivers are managed by various federal and state agencies.

Water Rights

The right to use the water in the river is a separate issue. Water laws refer to the rights of people, communities, and businesses that wish to withdraw and use water from flowing streams or groundwater. Water rights are one of the primary factors that influence floodplain management. Most state water rights systems allow the state to maintain a minimum instream flow for fish habitat and other public uses on each river.

Each state has developed, generally through a combination of legislation and the courts, unique determinations of water use and allocation. Those statutes should be consulted for specifics. As water becomes more scarce from increased development or climate change, individual states will have to wrestle with how to enforce water rights or make changes to their water laws. Neither choice will be easy.

Areas where rainfall, and thus water, is relatively abundant (typically east of the Mississippi River) generally use riparian water rights.

Riparian water rights associate the right to use water with the ownership of land beside or within which water flows.

Riparian owners are permitted to use all the water they need for their *proper purposes*, returning to the stream all that is not consumed, without liability to downstream riparian owners. Usually, what is proper are uses for households, farms, municipalities, and businesses.

The policy behind the law of riparian rights is to accommodate as many reasonable uses of a water resource as possible. Or, in other terms, to use the water as efficiently as possible. In most years there will be sufficient water to accommodate all users. However, in the event that the demand for water exceeds the supply (as during a drought), all users are expected to reduce their demand proportionately unless a reduction causes irreparable harm to a specific consumer (for example, if a person would be put out of business if denied sufficient water).

In most areas west of the Mississippi River, rainfall is relatively scarce. In arid regions (and increasingly in wetter regions) the demand for water exceeds the supply. Water rights then determine allocations. The western states generally use the doctrine of prior appropriation, also known as the Colorado Doctrine.

Under the doctrine of prior appropriation, water rights are established by actual use of the water and maintained by continued use and need. Beneficial purposes are usually defined as diversion for irrigation, mining and industrial applications, stock watering, domestic and municipal uses, and other non-wasteful economic activities. Water rights are similar to real property rights. They can be conveyed, mortgaged, and encumbered in the same manner, independently of the land on which the water originates or on which it is used.

In essence, the allocation rests upon the maxim *first in time, first in right*. The first person to use water (called the senior appropriator) acquires a right to its future use against later users (called junior appropriators). Because of scarcity, the number of junior appropriators is limited to that number for which water will be available after those senior have taken their allocations. In times of shortages, junior appropriators may not receive any water after senior appropriators have taken their entire allocations. Water rights may be forfeited if an appropriator does not divert water for a specified period of time, usually a period of years.

Property Rights

Land ownership rights are sometimes referred to as a bundle of sticks with each stick representing a right, such as the right to possess, sell, lease property, develop, mine ore, and so forth. But not all rights out of the bundle must be owned by the landowner. In the United States, no owner ever holds the fullest possible bundle. Even the federal government's ownership of land is restricted in some ways by state property law.

Traditionally, the bundle of rights includes:

- Controlling the use of the property.
- Benefiting from the property (for example, mining rights or rent).
- Transferring or selling the property.
- Excluding others from the property.

This bundle differs throughout the country. A property owner in Michigan may have the right to use water alongside the property in a reasonable manner, while a landowner in Idaho may be prevented from using water from a similarly situated stream because someone else owns the water.

Land can be owned by private citizens for private use or be owned by public institutions for the benefit of the public. Depending on the rights possessed, both private and public landowners are generally permitted to use land as they see fit. However, this bundle of rights can be added to and subtracted from with the adoption of specific laws and zoning.

What options are available to reduce flood hazards on your property? Generally speaking, so long as an action complies with building, land-use, and zoning laws, and the action does not adversely affect the property or rights of others, the owner is free to act. It is more difficult to control the actions of others, so you might have to buy their property or buy just the rights you need. Governments or government-sanctioned entities usually have greater leeway and can often purchase a wider range of rights, such as a conservation easement, or even transfer development rights between areas.

The purchase of property and property rights is a very powerful tool. After the 1993 Mississippi floods, FEMA provided money to enable local communities to purchase flood-prone properties. It reduced the community's flood risk and often, through providing storage on this cleared land, reduced the risk to downstream communities.

Transfer of development rights (TDR) has been used extensively for historic preservation in downtown areas or declared historic districts, but can be adapted for natural resource preservation. The owner of an existing historic property that is zoned for more intensive use is able to sell a development credit to another property owner. That owner can then use the credit to develop at higher densities than zoning would normally permit on his property. This transfer provides economic incentive as well as regulation to preserve the historic property. TDR programs usually require designation of *sending areas* (for example, a historic riverfront district) and *receiving areas* where increased development can be accommodated with needed infrastructure and services. The local government functions as a broker in TDR transactions and may maintain a TDR bank that can buy and sell development credits.

For natural resources, TDR programs are most commonly used to preserve open space or farmland in rural or rural-residential areas on the periphery of metropolitan areas. Such sending areas, usually within incorporated portions of counties, are paired with receiving areas in more urbanized portions of counties or with cities. Municipalities, though they may have their own TDR programs that operate internally, have the opportunity for regional cooperation by serving as receiving areas for development credits sent from a larger external area in the county or region. In 2008, King County, Washington, purchased the development rights from the Plum Creek Timber Company for 45,400 acres of the Green River upper watershed. In exchange, the timber company will be able to build more than 500 residences in King County. The purchased area will provide flood storage and drinking water for the Seattle/Tacoma metropolitan area and the transferred new development will take place in an already urban area.

Here are a few options for purchasing some or all private property rights to reduce hazards or enhance values.

- Public purchase of open space, development rights, or conservation easements.
- Purchasing land or development rights for constructing and managing drainage systems, for buyout of floodplain properties, or funding home elevation subsidy programs.
- Transfer of development rights (TDR) between property owners.
- Property owners donating conservation easements to a land trust in exchange for reduced taxation and other benefits.

Public Policies and Laws

The word government is from the Greek *kybernan*, which means to steer or to control. Government can be thought of as a way to steer laws and services so that our river-associated values are preserved, protected, or enhanced. It enforces laws that we have in some form collectively agreed to, and administers services that we have collectively agreed that we need and will pay for. All levels of government have a role in floodplain management.

Federal, State, and Local Governments

At both the federal and state levels the executive branch includes departments, agencies, bureaus, and offices. Local governments include city, town, village, civil township, county, parish, regional, and special purpose district levels, each of which may have a number of departments, depending on the size of the locality.

Despite the fanfare given to federal agencies, local government has the greater effect on our daily lives. Although there are a few large federal players in flood management like the USACE, the Natural Resources Conservation Service (NRCS), and the National Flood Insurance Program (NFIP), the federal role is limited. Most of what you need to do to protect your riverine values is available at the local level. Even when there is a federal mandate, local government most likely must enforce it. We cannot describe all the different structures and frameworks of government levels. What is important is to understand that there are different responsibilities and authorities for each layer.

The federal government has some overriding authority. Through the U.S. Constitution, statute, case law, charters, and so forth, it has defined the rights and responsibilities of state governments. Each state defines the size and nature of the next level such as counties, cities, towns, and so on. These local governments can then define the size and nature of smaller districts. Some states have given counties authority to create cities and towns within their jurisdictions; some states have not. Some counties can collect tax from city residents and dispense these funds, and some cannot.

There are more than 35,000 special districts in the United States. They generally focus on a single task, such as providing a utility or fire prevention. Some specifically address dikes. Many are water-delivery or wastewater-treatment districts, which give them a special interest in river management. They often cross city or county borders, so they can be useful when trying to look at an entire watershed. Hydrologist Warren Campbell analyzed 600 stormwater utilities and found that the vast majority base their assessment on the average number of square feet of impervious surface within the parcels in question. This type of assessment produces revenue based on need, since more impervious surface increases the amount of runoff. It also provides an incentive to use low-impact development techniques to reduce future taxes.

Cities are established communities represented by municipal general-purpose governments. Cities may or may not include utilities, fire districts, emergency management, schools, and other local services. In some places there are towns, townships, villages, and other governments that have less authority than full-service cities.

Counties (including townships, boroughs, and parishes) are generally larger than cities, although there are some city-county consolidations and some cities that stretch across more than one county. Counties generally have limited power within incorporated cities. The jurisdiction of counties is generally limited to unincorporated areas, so counties must plan around and in cooperation with incorporated cities and towns that may form a patchwork across large portions of a county. Many counties, especially in western states, include large areas of federal and state public lands or tribal lands over which they have little or no jurisdiction.

Because counties are generally larger than cities, they sometimes encompass smaller watersheds in their entirety and major portions of larger watersheds. Although some counties are largely urbanized, most counties outside the central portion of metropolitan areas have land in rural or natural resource–based uses. This gives some counties more options and greater flexibility in managing watersheds and floodplains.

In some places, watershed councils exist for the purpose of research, planning, and/or enhancement of streams on a watershed basis. These typically cut across local government (sometimes even state) jurisdictional lines. Check with your state, county, or city to see if one exists in your area.

Governments typically do long-term planning for an area or region. Flood-related policy can be an element of land-use plans, capital-improvement plans, and hazard-mitigation plans. Long-term planning is important for many reasons. It:

- Sets a direction that can be followed after a major disaster when opportunities for change exist.
- Involves the public in the process of looking at the future rather than just the present.
- Can produce a set of implementation measures that can be undertaken as resources become available, but still work toward a coordinated goal.

Several plans that address your values may already exist. These could include a city or county master plan, shoreline plan, hazard mitigation plan, park and recreation plan, or a river restoration plan. Although the plan itself may have been mandated by a federal or state agency, implementation usually rests with a local government. These plans are developed with advice from the public and hold valuable information on the intentions of those who prepared the plans. They may also have an inventory of existing resources and they often guide police powers and financing.

One of your first steps should be to go to your local government's Planning or Development Department and look at these documents. Plans are the product of a public process and they are usually updated on a regular basis. It is important to become involved in the planning process. Knowing what directions have already been discussed and deemed acceptable can provide political capabilities.

Local governments have primary authority to manage and regulate land use within their boundaries. The extent of this authority varies significantly from state to state. Most U.S. cities and counties develop comprehensive plans that include a future land-use element. They are adopted as policy and implemented through zoning and subdivision ordinances and other development codes.

Protecting the local tax base is another principal value of local government. However, this is more than just protecting government revenues. It relates to the more general value of supporting the economic health and development of a community. Public investment in transportation and other infrastructure and services is central to this objective. Reducing risk through emergency services, including all forms of hazard mitigation, also supports this purpose.

Increasingly important to economic development is articulating and promoting community image, sense of place, and amenities that attract business, new residents, and tourism. In most communities, these values are largely built on:

- Preserving and enhancing natural and scenic assets.
- Cultural, built, and historic resources.
- Cultural and educational institutions.

In the case of river communities, many of these assets are tied to the natural features and cultural artifacts of its river environment.

Laws

Rivers are important to many different interests so they typically have several layers of control. If you perform an action along a river, you'll have to be aware of all the laws that govern that action. Laws have many forms.

- *Statutes* are formal, written laws enacted by a legislative authority, ratified by the executive, and published. Typically, statutes command, prohibit, or declare policy.
- *Case law* is court decisions that interpret statutes, prior case law, regulations, agency guidance, and other legal authority.
- *Regulations* are legal restrictions created by government administrative agencies through rulemaking that provide for sanctions or fines. Federal regulations are published in the *Code of Federal Regulations*.
- *Mandates* are obligations by statute or administrative authority. They can be funded or unfunded. Unfunded mandates by the federal government can force the state or local government to find the means to pay for what is required. State governments sometimes force unfunded mandates on local governments.
- *Executive Orders* are issued by the president to help direct executive branch operations. Some orders have the force of law. Governors have similar authority in their states.
- *Guidance* can be offered by agencies operating under legal authorities. Often this guidance has the effect of law.

The Clean Water Act (CWA) demonstrates how all this fits together.

- *Statutes*: The Clean Water Act was passed into law in 1972 to control water pollution.
- *Case law*: In *United States v. Riverside Bayview Homes, Inc.*, 474 U.S. 121 (1985), the courts upheld the act's regulation of wetlands that intermingle with navigable waters.
- *Regulations*: EPA published effluent regulations and guidelines that regulate water pollution in fifty-six industry categories.
- *Mandates*: Municipalities were mandated to develop wastewater treatment plants to meet the standards set forth in the guidance.
- *Executive Orders*: In January of 2007, President Bush amended Executive Order 12866, further curbing the power of agencies such as EPA to offer guidance.
- *Guidance*: Section 404 of the Clean Water Act (CWA) regulates the discharge of dredged or fill material into waters of the United States, including wetlands. Guidance is the responsibility of EPA, but the day-to-day administration of required permits is through the USACE.

The National Environmental Policy Act (NEPA), signed on January 1, 1970, offers another model on how federal intent is implemented. This act offers a process to assess the environmental impacts of federal actions. States have adopted similar legislation and most state regulation offers substantive guidance as well as procedures.

The first time through the system may be confusing. The agencies and departments you need to talk to will depend on the type and scope of your project. Some cities have an ombudsman to assist. If you start at the top of the hierarchy, such as the USACE Section 404 permit officer, they will tell you what they will need and refer you to the next regulator.

Police Powers for Safety, Health, and Welfare

Police powers are a specific type of law that refer to the authority of government to enact and enforce laws and regulations protecting safety, health, and welfare. Although the term police powers may have an aura of heavy-handed regulation, there is a positive reason these powers exist. Land-use and other regulation are necessary to protect our resources and our way of life. These powers should not be seen as something to reluctantly use as a last resort. Instead, they are a way of enforcing policies that are necessary to allow our civilization to continue. We need an agreed-upon framework in which people can make rational decisions for the good of the community. Police powers allow that to happen.

The federal government does not have general police powers, though it certainly has great authority. The Commerce Clause (Article I, Section 8, Clause 3) of the U.S. Constitution empowers the United States Congress "to regulate Commerce . . . among the several States, and with the Indian Tribes." The Commerce Clause is one of the few powers specifically delegated to the federal government so its interpretation is very important in determining the scope of federal legislative power.

Still, most police power is granted to states, and through them, to local governments. The Tenth Amendment to the U.S. Constitution grants this power to the states. The federal government has the power to regulate only matters specifically delegated to it by the constitution. Other powers are reserved to the states or to the people.

Local governments tend to be the center of police power, which includes such regulatory tools as land-use planning, zoning, subdivision controls, and building codes. Land-use planning and zoning are specific police powers and can be used to limit and shape development on high-risk sites. If your goal is regulation, you'll find yourself entering the political arena. For example, assume upper watershed development is resulting in additional runoff that floods your property. You can work with your local government (city council, county commission, and so on) to control the development.

The NFIP has been the principal means by which local governments have regulated floodplain development. Most participating communities have adopted the minimum specified standards. However, communities have the ability to adopt higher standards. This could include requiring elevation above base flood elevation for new residential construction, adding channel migration zones to the regulated floodplain, and designating high velocity zones within the floodplain. These and other measures can also qualify a community for reduced flood insurance premiums under the Community Rating System (CRS).

Concern about *takings* is an issue often raised about land-use regulation. The Fifth Amendment says: "nor shall private property be taken for public use without just compensation." Supreme Court decisions have addressed this issue:

- *Pennsylvania Coal Company v. Mahon* (1922) stated that a government regulation might restrict the owner's freedom to use his property to such an extent that it can constitute a taking of that property without compensation.
- In *Kelo v. City of New London* (2005), involving the transfer of land from one private owner to another to further economic development, the court held that the general benefits a community enjoyed from economic growth made such redevelopment plans a permissible public use under the takings clause.
- The right of local governments to enforce their laws was clarified in *Village of Euclid, Ohio v. Ambler Realty Co.* (1926). The Village of Euclid wanted to maintain their village character and the Ambler Realty Co. wanted to develop industrial uses. The village was relying on the relatively new practice of zoning to regulate land uses. The court argued that the zoning ordinance was not an unreasonable extension of the village's police power and did not have the character of arbitrary fiat, so it was not unconstitutional. Further, the court found that Ambler Realty had offered no evidence that the ordinance had any effect on the value of the property in question, but based their assertions of depreciation only on speculation. The court ruled that speculation was not a valid basis for a claim of takings.

There are two concerns with police powers. One is the threat of being interpreted as a taking. The other is that they are effective only if there is the potential for change. Police powers can protect communities against adverse effects of future development, but will have little effect if the development has already occurred.

Localities have historically been reluctant to use zoning to limit floodplain development because of takings issues. Courts have struck down zoning ordinances that prohibited complete economic use of a site, but they have tended to support zoning based on health, safety, and welfare arguments when some economic use was permitted. Cluster zoning, including planned unit development, concentrates development on a portion of a site, leaving open space available for hydrologic functions, recreation, and other purposes. This generally avoids takings issues, permits development at existing or higher overall densities, and is often highly marketable, based on amenities that can be incorporated. Subdivision ordinances can also be effective tools for shaping development.

Land-use zoning, subdivision, or building codes and ordinances offer a range of opportunities to reduce adverse effects and manipulate hazards.

- The NFIP uses local governments' police powers to reduce the effects of floodwater-related hazards. If communities do not agree to adopt and enforce NFIP zoning and subdivision requirements they risk being sanctioned and flood insurance will not be available to their residents.

- Agricultural or Forest zones set a minimum lot acreage for land that must be dedicated to farm or forest operations. Typically, little or none of the land can be used for structures. This type of zoning can dedicate watershed lands for rural uses. Both agriculture and forestry can disrupt natural ecosystems and streamflow patterns, but not nearly to the extent that urbanization changes the character of the land and water.

- Cluster or planned unit development (PUD) zoning and subdivision ordinances provide some land use planning tools to implement a low-impact development pattern. Planned unit development typically permits clustering of lots and buildings at higher densities on designated portions of a large development site in exchange for open space set-asides for identified functions and purposes. The result is the entire PUD has the same average density specified by underlying zoning.

- Subdivision regulations can be applied to a PUD or conventional development to control the configuration of buildable lots and delineate specific tracts to remain in open space. For example, open space tracts can be designated at the heads of natural drainage or stream channels where gradients are steepest. Keeping natural vegetation can minimize acceleration of runoff at these points.

- Practicing low-impact development (LID) methods can reduce flooding downstream and increase onsite water storage. LID can be encouraged by providing information on the approach, offering density or tax bonuses for their construction, paying for the stored water, or adding the stored water recharging the aquifer to a water market.

- An urban growth boundary (UGB) can be established to contain an urban area. A perimeter is designated and urban development is encouraged within the perimeter and discouraged beyond it. Typical tools include high density within the perimeter and resource use or very low density residential development outside it, and not allowing the extension of urban services outside the perimeter. However, care must be taken not to make the UGB so large and of such high density that it unalterably changes regional watersheds and creates significant problems downstream.

- Building codes set standards and requirements for the construction, maintenance, operation, occupancy, use, or appearance of buildings, premises, and dwelling units. They offer an effective way to ensure that development is built to withstand natural hazards.

Funding Opportunities

Local taxes can provide incentives (and disincentives) to reduce risk and achieve public benefits. These include:

- Using impact fees, special assessments, or surface water utility charges to generate revenue to purchase land or development rights; for construction and management of drainage systems; for buyout of floodplain properties; or funding home elevation subsidy programs. This can include special assessment districts related to particular hazards, such as a flood hazard management district where an added property tax assessment is applied for flood hazard reduction measures.

- Reducing property taxes for open space set-asides, based on hydrologic function and public benefit. For example, King County, Washington, adopted a stormwater utility that collects revenue to control runoff. They are also taxing land below market value if owners maintain a use that does not alter current runoff characteristics.

Federal money is sometimes available to help you protect, conserve, or enhance your values. This money is generally available only if there is a group who is at risk or will benefit, not a single property owner. Ferreting out sources may not be easy. Below are some programs as of 2009 that may offer a starting point. Many funding programs change quickly, as do their eligibility requirements and funding levels, so you will need to research current opportunities for your specific project.

Appendix A contains an explanation of the NFIP and some FEMA programs. Some NFIP-related programs can be used to fund flood mitigation. Here are five programs restricted to communities in good standing with the NFIP. There may be similar types of state funds available.

- The Hazard Mitigation Grant Program (HMGP) provides project and/or planning grants to states and local governments. Projects must be consistent with an adopted local mitigation plan and can include buying flood-prone structures, elevating structures, and using floodwalls to isolate vulnerable structures. Funds are available only after a federally declared disaster. There is a 75 percent federal and 25 percent nonfederal cost share.
- The Pre-Disaster Mitigation Program (PDM) provides project and/or planning grants to states and local governments. Projects must be consistent with an adopted local mitigation plan. There is a 75 percent federal and 25 percent nonfederal cost share.
- The Flood Mitigation Assistance Program (FMA) funds projects that are consistent with an adopted flood mitigation plan. There is a 75 percent federal and 25 percent nonfederal cost share.
- The Repetitive Flood Claims Program (RFC) focuses solely on reducing or eliminating flood damage to structures with many flood insurance claims. There must be a local plan. The federal government pays 100 percent of costs.
- The Severe Repetitive Loss (SRL) Program focuses on reducing or eliminating flood damage to residential structures that are a drain on the National Flood Insurance Fund. There is a 75 percent federal and 25 percent nonfederal cost share.

The agencies responsible for the bigger, mostly structural, projects are primarily the U.S. Army Corps of Engineers, the Bureau of Reclamation, and the Bureau of Land Management. These federal agencies respond only to official requests from governments. If you have defined your values in such a way that big projects need to be considered, then you'll need to engage the federal government. Because local matching funds from state or local governments are usually required, you'll also need to engage them.

There are also hazard-reducing funds from other agencies. For instance, the U.S. Department of Agriculture (USDA) Natural Resources Conservation Service (NRCS) has a suite of programs available to ranchers and farmers for holding water in the higher watershed, thereby conserving flows and reducing soil erosion and downstream flooding. The USDA Forest Service has similar programs directed to forest harvesters.

Integrating Capabilities and Tools

Many of the tools at your disposal will depend on your specific situation and capabilities. Such things as the willingness of the community to use zoning to restrict or deny floodplain development, eligibility for USACE assistance, or the existence of a watershed council that researches riverine and watershed conditions can make it easier to propose and implement new floodplain management systems.

Other capabilities are broader in scope. For example, an overall cut in USACE's budget may be something you have no control over. On the other hand, one of the capabilities is timing, and that can change at a national level. Trying to manage hazards, particularly floods, is not a new task and there are many lessons to be learned from the past. Recent events show that we may be entering a period when more attention will be paid to protecting property and especially lives from serious flooding.

- Images of desperate people in the wake of Hurricane Katrina show how devastating flooding can be. Even though many warned of impending levee failure, governments were not able—in some cases not willing—to act before the crisis. Some individuals foolishly believed they were not truly in danger, but others wanted to leave and had no way of evacuating the city before the hurricane struck and the levees broke. Katrina spotlighted the problem of not adequately providing for floods and other natural hazards in land-use and development plans. Much of the flooded area was home to the poorest of the city. The trade-off for affordable housing cannot be jeopardizing the lives of the people who live in it.
- The Mississippi is not the only waterway protected by levees. In 1986, a 150-foot gap opened in a Yuba County, California, levee, flooding hundreds of homes and a shopping center. The subsequent

lawsuit against the local reclamation district and the state was known as the *Paterno* case. The 2003 decision forced the state of California to pay nearly half a billion dollars to compensate property owners for the failure of levees. This case shows that courts are not willing to force victims to pay for poor development decisions.

- FEMA is now working to digitize and update all the Flood Insurance Rate Maps in the country through the Map Modernization program. As part of this effort, some state and local hazards managers are conducting studies that show dramatic increases in flood heights if floodplains continue to be developed. Those studies must be shared with all potentially affected communities and FEMA so that better decisions on future development can be made. Communities should partner with FEMA to produce more useful maps that contain flood hazards based on future conditions.

Attorney Edward Thomas, writing in the *Natural Hazards Observer*, describes how these and other events should affect the conversation about public policy.

> The concept of No Adverse Impact and the legal foundation it is built on can help develop win-win relationships between hazards managers and community development officials, developers, emergency managers, wetland managers, water quality managers, stormwater managers, and others. . . . Hazards-based regulations are generally sustained against constitutional challenges, and the goal of protecting the public is afforded enormous deference by the courts. By providing our local officials with a better understanding of the laws that affect them, especially those that are nonissues, we as hazards managers can help them diminish or prevent the misery caused by improper development. Our ability to supply this sort of information, to get involved, in a post-Katrina world of heightened awareness of natural hazards, should give hazards managers a welcome place at the table as development decisions are made. Nevertheless, we must aggressively seek that place at the table, and we must act fast.

Case Study: Davenport, Iowa

Davenport, Iowa, lies along the Mississippi River and is no stranger to floods. The city began looking at a floodwall in the 1960s. In 1970,

Congress authorized a levee project and in 1983 the USACE estimated the cost at $34 million. Davenport declined, partly for financial reasons, partly because the levee would separate the city from its river.

The decision was not popular everywhere. Floods in 1993, 1997, and 2001 damaged property in the area and some thought that Davenport should not receive federal disaster assistance because it had not done enough to help itself. One of the city's main flood management strategies during this time was to buy buildings in the floodplain. That removed some vulnerable assets and left the water room to pond during flooding.

In 2002, the USACE updated its estimate for a levee to $55 million. Because the city had removed so many structures in the floodplain, the levee no longer made economic sense.

1. What Values or Assets Do You Want to Protect or Enhance?

Davenport is a river city, with the history and amenities that follow from that. It wanted to keep its economic, ecological, and social ties to the river but reduce damage from floodwaters.

2. What Are the Apparent Risks or Opportunities for Enhancement?

Flooding was a major drain on the resources of the city, businesses, and individual residents.

3. What Is the Range of Risk-Reduction or Opportunity-Enhancement Strategies Available?

- The traditional approach was to build dams and/or levees. The USACE drafted a proposal for a levee.
- Less invasive approaches, such as removing buildings that could be damaged and restoring parts of the floodplain, also reduce flooding.

4. How Well Does Each Strategy Reduce the Risk or Enhance the Resource?

- Levees can dramatically reduce flooding, though they need to be built higher over time to compensate for the river's actions. A levee break can produce catastrophic flooding damage.

- Less invasive approaches can reduce flood heights and can remove assets from being damaged.

5. What Other Risks or Benefits Does Each Strategy Introduce?

- Levees have high maintenance costs and cut the city off from the river. They reduce plant and animal life along and in the river and eliminate aesthetic values.
- Nonstructural approaches do not eliminate flooding but can reduce its frequency and contain it, reducing the threat to human safety and property. Development along the river can be done so that assets that are periodically flooded are not greatly damaged (such as parks). This strategy also allows the community to stay connected physically and socially to its river.

6. Are the Costs Imposed by Each Strategy Too High?

- Levees have expensive long-term costs and can add to economic damage by reducing riverine views and degrading the wildlife resources of an area.
- Nonstructural approaches can have high initial costs but much lower maintenance costs and can enhance many intangibles of a community.

7. The Decision

Davenport declined to build a lengthy floodwall, an option more and more cities are choosing. The city built a park along the river, which floods but is easy to clean up. An art museum built in the area sits atop a parking garage, keeping it dry during flooding. The city is seeking to expand its riverfront in similar ways, to preserve community values and enhance economic development.

The city experienced major flooding in 2008, as did much of the upper Mississippi. However, many levees in other places were overtopped or failed in that event, clearly illustrating that levees are not an absolute protection.

Chapter 8

Strategies: Work with, Not against, Rivers

Approaches and tools can be packaged into strategies to reduce flood damage and enhance flood benefits. In the past, most flood-related strategies tried to get water off the land as quickly as possible. Now the benefits of keeping water on the land longer are becoming evident. Preserving old stands of trees, natural swales, wetlands, oxbows, vegetated riparian areas, and meanders; building small weirs, sediment structures, wet gardens, and setback levees; lengthening streams; and accepting large woody debris in the channel are useful techniques. Using these practices to work with the natural motion of the river is more effective—and less expensive—in reducing flood damage and helping us realize benefits from the water.

Flood Management Strategies

In this book we define a strategy as a package of approaches (described in Chapter 6) and tools (described in Chapter 7) applied over time to achieve an objective. In this chapter we will combine information about natural riverine systems, human uses, approaches, and tools to discuss what options are available to a variety of stakeholders.

In the past, most flood-related strategies were attempts to make our streams more efficient, with efficient defined as getting water off the land as quickly as possible. As water continues to become more valuable, we see greater benefits in keeping water on the land longer. This water spreading philosophy calls for creating areas of storage in our watersheds, floodplains, and channels.

We are also recognizing that there is no silver bullet—no single course of action that fixes all situations. A single large dam will not solve all our water problems because it creates many unwanted secondary effects. In most cases it is less expensive over the long term to work with the river rather than against it. It is vital to consider a range of strategies that address the risk you have identified. Too often in the past we started with the answer (dam, levee, and so on) and applied it to whatever question came up. Sometimes it is better to do nothing at the present time than to do the wrong thing and make the problem worse in the future. The costs of all available alternatives, including delay, must be balanced.

Strategies include:

- Solutions that restore, conserve, or enhance our values.
- Science-driven policy, not policy-driven science.
- A package of approaches that allows for flexibility. The situation (river, watershed, built environment) is always changing. There are always risks and opportunities.
- Solutions that promote resiliency, accommodating change over the long term.
- A system-oriented outlook—inputs, attributes, and outputs—based on the watershed as a whole. For example, the solution to a floodplain problem may lie in the higher watershed, far from the active floodplain.
- A process that involves all stakeholders.
- Solutions that, over the long term, offer greater benefit than risk to the river and its stakeholders.
- Equitable solutions minimizing adverse effects on all stakeholders and the environment.

Megastrategies

We offer a set of generalized values on which to base a national set of strategies to reduce flood risks and enhance their benefits. At the broadest level, there are three—the ability to:

- Live and work in an area that is not adversely affected by flooding.
- Have clean water available.
- Enjoy river amenities.

The physical and biological changes driving risks to these values include:

- Climate change and associated changes in the amount of rain and seasonal runoff will change runoff and erosion patterns. More extreme weather is likely, as seen in the increased number of heavy rain events in the United States.
- Changes in erosion patterns will affect plant and animal communities and pollution loads, which may degrade river conditions.
- Increased urbanization contributes to less storage, higher and more frequent floods, more soil erosion, and less evapotranspiration, resulting in reduced soil moisture and greater fluctuations in river discharges.
- Existing levees and dams have changed the complex biota of plants and animals that naturally dampen flood heights and control water quality.

Societal changes include:

- Sprawl and consumption of flat land associated with floodplains and rivers. Population and development are consuming our flatland resource.
- National Flood Insurance Program (NFIP)–inspired floodplain development is reducing the effect of flooding on structures by elevating them. The elevation, however, is often done through fill, which reduces storage, increases river discharges downstream, and can degrade water quality, creating a hostile environment for plants and animals.
- The NFIP, along with the No Adverse Impact (NAI) initiative and the Community Rating System (CRS), is now developing a culture of accepting floods. Such a culture provides the foundation to embrace low-impact development (LID) approaches that use storage as a way to reduce discharges and lower flood peaks.

Past approaches and tools have often resulted in increased flood risks over the long term. These risks appear to be increasing with urbanization and climate change.

- The principal historical approach to flooding has been to increase discharge efficiency through channel straightening, levees, and dredging. This often exacerbates downstream flooding.
- Very large structural projects often have a short-term economic benefit but represent an increased risk over the long term.
- Property owners often have the right to maximize the development of their property, including impervious surfaces, without regard to the

cumulative harmful effects on others. This contributes to increased river discharges and loss of basin storage.

- Urban stormwater systems have substantially changed runoff patterns in many cities.
- Often floodplains are managed as if they begin and end at government jurisdictional limits. There are few examples of basin-wide comprehensive management authorities.

We also have, however, many opportunities to improve our river and watershed management.

- Older structures are continually being replaced and new development is always occurring. This can provide opportunities for a better built environment using LID and NAI principles. Tools such as police powers, the purchase of development rights, and development clustering can be designed to accommodate floods, be sustainable, and take advantage of natural beneficial systems with less risk.
- Opportunities that take advantage of natural systems to reduce adverse flood effects and store water have not yet been generally accepted. Stream restoration activities that increase in-channel and floodplain complexity also provide flood abatement benefits. In the past, many floodplain managers have attempted to incorporate "natural and beneficial" values almost as an afterthought to river projects.

At the most general level, a national strategy could pair leadership with financial incentives and guidance to support the values mentioned above. Specific steps could include:

- Floodplain managers will search for solutions within the larger watershed, not just on floodplains.
- In upper watersheds, we can conserve, restore, or enhance the ability of the landscape to filter, store water, and temper discharge rates by taking advantage of better forestry and agricultural practices.
- In the middle of urbanizing watersheds, we can conserve, restore, or enhance the ability of the landscape to process pollutants, filter, store water, and temper discharge rates by concentrating development away from critical storage areas (defined by both the physical and biological landscape) and requiring LID and NAI development techniques.
- Along the river corridor we can allow the river to lengthen by using setback levees and take advantage of natural processes to filter, store water, and temper discharges. Structural improvements, land devel-

opment, and other human practices should be designed to accept the river's natural, meandering corridors.

- Not all river reaches are equal. Where possible, development should be kept off the floodplain. Strategic areas can be identified where development must be avoided. Some of these are high-energy reaches, areas where complexity is needed to supplement otherwise homogeneous reaches, and opportunities to promote connectivity to restore environments.
- Levees and dams should be removed or modified when practical to allow some river reaches and watersheds to return to a more natural state.
- If levees are the only practical solution, opposite banks should not be hardened over extensive lengths. This allows the river to maintain some natural characteristics.
- Organizations addressing floodplain issues must promote a cultural understanding and acceptance of flooding and the associated benefits.
- Tools can be developed that recognize property rights but still address cumulative adverse actions by encouraging winners to reimburse losers.
- Basin-wide management authorities must be created and supported. Existing examples include the Tennessee Valley Authority and the Great Lakes Water Resources Development Act, along with other local, state, and interstate watershed councils.

There will be trade-offs in revising previous strategies to reduce flood damage. The idea of exploiting natural processes to reduce the adverse effects of flooding while enhancing the positive ones often flies in the face of our cultural wisdom, making acceptance difficult. We have a history of thinking we have the engineering expertise to redesign rivers to be more efficient or to be fixed. Over the long run, we don't.

With the increasing need for water, big money and big interests will likely recommend fast, expensive fixes—expect recommendations for large dams, major watercourse diversions, and channel straightening. Such large projects have often initiated a cycle of degradation and increased maintenance costs.

Strategy by Stakeholder

Many strategies are suitable for a variety of stakeholders to use. In general, each larger level of stakeholder can also use strategies described for a smaller one. Neighborhoods can use the tools available to individual property owners, as well as some additional ones. Local governments incorporate both of those sets, as well as their own authorities. States have even more capabilities.

Individual Person

The river-related values of individual property owners are typically to protect structures from flood damage, realize a profit when they sell, and enjoy river amenities. They may not seem to have many capabilities. However, a single person can have an enormous effect by becoming a champion for new flood management strategies. Energizing a community, working with others, and lobbying for change have been used successfully through the years in many policy areas.

Wangari Maathai founded the Green Belt Movement in Kenya in 1977. It is a grassroots environmental, nongovernmental organization that has planted more than 30 million trees across Kenya to prevent hazards such as flooding and soil erosion and to provide a sustainable fuel source. Maathai and the Green Belt Movement are credited with saving Nairobi's Uhuru Park in 1989. In the early years of the organization, Maathai was imprisoned and violently attacked for her work. She won the Nobel Peace Prize in 2004.

For individuals, risks usually involve water damage to something they value. High water, sediment, and channel migration can be threats, as can the hidden risk of being sued for doing something on the property that would cause an adverse effect on someone.

Physical and biological conditions govern capabilities. Slow-rising shallow flood stages allow property owners to live with occasional flooding and use elevated homes as a solution. High velocity flows, high stages, and flash floods may greatly limit that alternative. Silt-laden flows may require approaches that help distribute the material. Impermeable soils or owners' lack of interest may limit opportunities for wet gardens.

Altering the mechanics of the disturbance is difficult and includes projects like building levees. As previously discussed, levees can change how often floodwater reaches a structure and alter the warning time and even the location of the floodwater, but these changes in frequency are usually at the expense of proportionally increasing flood damage. Levees are expensive to maintain, often costing more than the protected land is worth, though maintenance costs are often borne by a larger community, reducing the cost to riverside property owners.

The NFIP provides incentives for reducing the harmful effects of flooding but does not address disturbances such as channel migration, erosion, sediment deposition, climate change, or future development. These processes often require addressing the problem within an entire watershed. If they occur and are severe, the homeowner might decide to leave that site, possibly through a government buyout.

All approaches have trade-offs. Elevating a structure can reduce flood damage. If fill is used to accomplish this, however, it can increase runoff, which can increase flood damage downstream, resulting in liability issues. And, even though the structure is protected, the family will not be able to inhabit their house when it is inundated by floodwaters.

Neighborhoods

A neighborhood in a floodplain or near a river has strategies for collective action that can protect assets and enhance values. This can be done through private, cooperative efforts and through influence brought to bear on local government.

Values of neighborhoods include:

- Reducing risks that are common to the community as a whole.
- Securing affordable flood insurance.
- Maintaining property values.
- Avoiding land-use conflicts.
- Maintaining property rights and the fair resolution of conflicting individual or private-versus-public property rights.
- Enhancing neighborhood infrastructure, public facilities and services, and amenities.

Where a neighborhood faces risks to property from flooding, erosion, channel migration, or other riverine disturbances, collective solutions tend to be less costly, more effective, and less likely to produce conflict than individual landowner actions. Clearly to be avoided are actions where an individual landowner introduces flood protection on his property that diverts floodwater to neighboring properties. Fill, elevating a building with a closed ground story, or diking can do this.

However, when owners of adjoining properties work together, the options for managing water and instituting LID provisions are greatly expanded. For example, temporary storage of runoff or floodwater is difficult to accomplish on individual half-acre to 5-acre lots in suburban or rural areas. In most cases, too much of each parcel is devoted to buildings, driveways, or developed outdoor activity areas. But subdivisions often permit the creation of larger, common storage areas that run across lot lines. These may include depressions, minor drainages, or wetlands.

Many newer subdivision ordinances in urban areas require that open space tracts be designated for purposes of detention and drainage, though in most cases they do not require sufficient storage. In most rural residential areas, as well as older suburban subdivisions, no such provisions exist. However, property owners can achieve the same results by working together. Also, when further land division occurs, applicants are well advised to work with neighbors as well as their local government permitting authority. Together they can find the best solution, resulting in economic returns while protecting neighborhood assets and amenities.

Common stormwater management by a group of adjoining homeowners can include features such as wet gardens and swale systems that capture and slow runoff, routing it across multiple lots to natural drainage ways. This is even effective if it replaces existing storm drains.

Other low-impact approaches such as green roofs may be confined to individual properties but neighbors can pool their expertise, reduce costs through the bulk purchase of materials, and participate in other forms of cooperation.

Managing vegetation, including neighborhood tree planting and maintaining ground cover, is another activity conducive to neighborhood volunteer projects. Many areas of the country go further and offer adopt-a-stream opportunities to clean up streams and restore habitat.

Neighborhoods can also exercise significant political power and influence with their local governments and special districts. In some cases, local governments establish neighborhood or community councils. In other places, citizen concerns about floodplain risks and benefits can provide the initial basis for community organization.

Through collective action, citizens can be strong advocates for effective floodplain management in their communities. They can, for example, watchdog their locality's implementation of floodplain development regulations under the NFIP and push for additional measures that would lower homeowner flood insurance rates through the CRS.

Designation or non-designation of floodways is another aspect of floodplain regulation that typically is responsive to public influence. Where purely conveyance-oriented approaches threaten community values (including, in some cases, development opportunities), citizen intervention can make a difference.

Citizen groups are influential in planning for both land-use and capital improvements. Local governments generally apply uniform standards throughout their jurisdictions for streets, sidewalks, bridges, storm drains, and other infrastructure. Residents of neighborhoods can insist on flexibility for localized conditions and preferences. Excessive, uninterrupted fills for roadway berms or street grading, excessive street widths and impervious surfaces, and excessive use of piped drains and storm sewers as opposed to surface management of runoff are all examples of infrastructure based on rigid standards. They may not address local conditions, concerns, and values—or area-wide hydrologic performance. The typical six-year cycle for capital improvements provides an opportunity for community participation.

Neighborhood citizen groups can be effective advocates for preserving or adding community amenities. This can be accomplished in part by influencing local government investment in parks and recreation facilities, riverfront access, protection of habitat and other open space through purchase of land or development rights, acquisition of historic sites, providing incentives to private landowners to preserve natural and cultural resources, and other options.

Many places allow residents to form a local improvement district, which can assess an additional, small increment of property tax. The revenue can either directly pay for specified improvements in the designated area or retire bonds issued by the local government to finance improvements. These districts can do a variety of tasks from acquiring neighborhood parks and open space to improving surface water management.

Local Governments

The authority given to special districts, cities, and counties depends on specific state laws and varies across the country. You'll need to research what local government jurisdiction is appropriate for action in your area.

We assume here that local government, as a stakeholder, is representative of interests within its borders. Most larger river cities and many smaller cities and counties in riverfront or floodplain locations are mostly urbanized. This presents some limitations and challenges for developing strategies to manage risk and realize benefits.

Cities, counties, and special districts have an obligation to function as responsible public entities within a larger region and watershed. Many cities have erected structural flood protection like extensive river-fronting levees and floodwalls, causing what may literally be described as spillover costs on neighboring communities. Even so, it is unreasonable to oppose all existing forms of structural mitigation, especially in highly urbanized situations. Each case must be evaluated on its merits from the standpoint of risks and benefits to all affected stakeholders. Some economically viable and historic downtown areas are currently protected by a floodwall. It may make economic sense to preserve these structures until better area-wide solutions can be found. This is especially true if the effect of the floodwall on surrounding areas has been manageable or capable of mitigation in a manner fair to affected property owners and other stakeholders.

FEMA floodways are strictly passageways for the conveyance of floodwaters. Local governments can go beyond FEMA minimum requirements. Floodways could be used as open space corridors that, depending on conditions, can provide conveyance or storage—or both. Some states require local governments to delineate environmentally critical areas such as wetlands and flood-prone areas and apply more stringent land-use regulation in these areas. This can provide a legal basis for floodplain measures that go beyond NFIP requirements.

Local governments can also offer incentives to landowners to protect community assets and reduce their hydrologic footprint. Some incentives work directly in combination with regulation. Incentives can be offered to builders who use cluster zoning and LID practices. Conservation easements dedicated to a city or nonprofit conservancy offer landowners a substantial package of tax incentives while still allowing some development on a site. Preserving historic urban fabric along riverfronts and adjacent districts can also be supported with tax incentives, partial redevelopment options, and assistance programs.

Transfer of development rights (TDR) is another incentive-based approach. Cities can serve as receiving areas for development credits sent from another area in the county or region. In this way, cities can be good neighbors in helping to address floodplain and watershed issues at a regional scale.

Much of the thrust of contemporary growth management is to increase density in existing urban areas as a means of preserving farmland, forest, and other open space on the periphery. There are limits, however. At some point, high urban densities begin to preclude any options for returning an area of land or a section of river to a more natural hydrology. Local governments should establish their objectives and plans for floodplain management first, and then consider land-use options and development opportunities that are consistent with those plans.

Local governments can acquire land or development rights to land through the exercise of eminent domain or through negotiated purchase.

Many have purchased extensive areas of land along their rivers and preserved it as open space. Open-space bond issues have been widely successful as financing mechanisms. Federal and state conservation programs have provided some funding for localities to purchase. Many communities have obtained funding under the federal Land and Water Conservation Fund to acquire land for park and recreation areas.

A local government can use a coordinated strategy to reduce risk and reduce the hydrologic footprint of a community and its effect on neighboring areas. It can also enhance its river-related amenities and sense of place. Over time, such a strategy can allow even an established, largely built-up urban area to step back from a purely conveyance approach, replacing it with one that gives back some land to the river; widens its cross section; restores vegetated soft edges, riparian areas, and back channels; and allows for some on-land storage during floods. Such a strategy can be combined with the preservation or creation of urban open space and its development for the recreation uses that are increasingly in demand in urban areas.

Counties have a particular opportunity in playing a regional coordinating role. In some cases, state growth management programs require coordinated land-use planning between counties and their constituent cities, or even among counties in the case of a few metropolitan areas. For example, the Washington State Growth Management Act requires that the four central Puget Sound counties adopt a coordinated growth strategy and designate urban growth area boundaries together with their municipalities. The act then requires that different zoning and development standards be applied to urban areas within these boundaries than to rural areas outside the boundaries. In 2009, the legislature strengthened the law by prohibiting western Washington cities from extending their urban growth areas into 100-year floodplains. Even when not required, however, counties can play an important role in coordination.

Local governments can conduct hydrologic analyses of watersheds and develop management plans for entire floodplain areas. Some counties, such as Snohomish and King counties in Washington, have been quite successful in limiting floodplain development and maintaining rural land uses through as basic a tool as agricultural zoning. Cluster zoning is often most effective in a rural situation involving larger blocks of land where development can be concentrated outside of high-risk zones or outside of the floodplain.

Local governments may be able to invest in and manage infrastructure at a regional level. Often this includes managing levee systems along significant portions of rivers. It may also include managing dike systems extending across floodplains, though in some areas these systems are managed by local diking districts. Although the U.S. Army Corps of Engineers (USACE) originally constructed many existing levees along major navigable rivers, much of the management and nearly all repair and new construction fall to local governments.

Local government strategies could include:

- TDR programs, either county-based or interjurisdictional.
- Regional agreements to coordinate land-use planning and development standards, and to agree on future annexations by cities.
- Farmland preservation programs involving government purchase of development rights along with produce marketing and other assistance to farmers.
- Extensive use of conservation easements, often in partnership with local or regional land conservancies.
- Working with private, state, federal, and other entities to create scenic and recreational greenways along river corridors.
- Setting back levees to allow a river to recapture a portion of its floodplain and restore some physical and biological processes.
- Modifying levees to allow overtopping at flood-tolerant locations.
- Reconnecting back channels and wetlands to the river.
- Creating passageways for flow under roadway berms.
- Establishing temporary or seasonal storage areas through purchase or agreements with landowners.
- Adopting LID methods that increase water retention and reduce dependence on engineered, below-ground, piped stormwater detention and conveyance.
- Requiring that post-development runoff not exceed predevelopment levels.
- Increasing detention opportunities within and off the floodplain through police powers or the outright purchase of flood easements and other development rights.

State and Federal Governments

State and federal governments have different authorities but similar values. Both want to reduce flood damages to their constituents and assets. Both want to maintain a clean, viable water resource.

A comprehensive review of all state programs is beyond the scope of this book, but one example is the state of Washington's Flood Control

Assistance Account Program (FCAAP). This program was created in the 1980s to help local governments and tribes reduce flood hazards and damages. Administered by the Department of Ecology, FCAAP provides technical and financial assistance to develop and implement comprehensive flood hazard management plans, engineering feasibility studies, physical flood damage reduction projects, public awareness programs, flood warning systems, and emergency projects; and to acquire flood-prone properties. Matching grants are awarded and coordinated with the Washington Emergency Management Division's administration of FEMA grant programs. The guiding principle for grant awards is a management approach that maximizes coordination between providing public safety, protecting public infrastructure and private property, and preserving natural resources.

Since its inception, FCAAP has provided nearly $40 million to communities throughout the state. Two successful projects were:

- Kitsap County's Clear Creek floodplain restoration project, which acquired flood-prone properties and restored severely altered natural conditions. This substantially reduced flooding problems and enhanced the ecological functions of the area.
- Pierce County's upper Puyallup River project, in partnership with the HMGP, which helped acquire flood-prone properties and remove nearly 2 miles of levees. This created significant flood storage, restored valuable habitat, and reduced the severity of flooding in downstream communities. It also encouraged the county to initiate a levee setback program that continues to yield additional benefits.

The federal government is restricted to addressing their assets, such as national parks, national forests and grasslands, interstate highways, and those resources that fall under the Commerce Clause such as rivers that cross state boundaries.

States have authority over everything else so long as they don't violate the U.S. Constitution, for example by taking property without due compensation. However, in most cases, they are only tangentially addressing rivers and floodplains within the context of watersheds. There is little acknowledgment within state and federal legislation of the interdependence between forest and agricultural harvest practices, impermeable land cover, and climate changes and their effects on flooding, water losses, and drought.

With tens of thousands of separate authorities, coordination at the federal and state levels is needed. Until the early 1980s, coordination at the federal level was supported by the U.S. Water Resources Council (WRC) and aided by their *Principles and Standards*.

These were replaced in 1983 by the *Economic and Environmental Principles and Guidelines for Water and Related Land Resources Implementation Studies*. In adopting these *Principles and Guidelines*, James G. Watt, chairman of the Water Resources Council, said he was "Confident that this new guidance . . . [would] . . . enhance our ability to identify and recommend to the Congress economically and environmentally sound water project alternatives."

WRC funding and most responsibilities were terminated under the Reagan administration, with the exception of responsibilities for the Interagency Floodplain Management Task Force that were transferred to FEMA along with one staff person. FEMA continued to convene meetings of the task force until the late 1990s. As of 2009, there is no formal structure to coordinate water resources policy.

FEMA has not integrated the *Principles and Guidelines* into their project guidance. However, the USACE Civil Works, Bureau of Reclamation Water Resources, Tennessee Valley Authority Water Resources, and the Natural Resources Conservation Service (NRCS) Water Resources use the *Principles and Guidelines* to assess projects. Other federal agencies and states are encouraged to include themselves in the process.

This guidance established four accounts designed to evaluate projects that affect water resources.

- The National Economic Development (NED) account displays changes in the economic value of the national output of goods and services.
- The Environmental Quality (EQ) account displays nonmonetary effects on significant natural and cultural resources.
- The Regional Economic Development (RED) account registers changes in the distribution of regional economic activity that result from each alternative plan. Evaluations of regional effects are to be carried out using nationally consistent projections of income, employment, output, and population.
- The Other Social Effects (OSE) account registers plan effects from perspectives that are relevant to the planning process but are not reflected in the other three accounts.

The *Principles and Guidelines* are procedural and implemented on a project-by-project basis. Unlike the National Environmental Policy Act

(NEPA), they are economically driven and NED accounts take precedence over other planning elements. Also unlike NEPA, they have no clear agency oversight.

Probably the *Principles and Guidelines* guidance's fundamental flaw is that it assumes that the four accounts are unique, can be addressed separately, and that a balance is possible. We disagree. This book promotes reducing adverse effects through an integration of natural processes and assumes that our built environment is not separate and apart from the natural world.

The *Principles and Guidelines* leave a dire need for nationwide coordination. This need will increase with expected changes in land cover and climate.

Here are a few examples of existing federal water management strategies:

- The NFIP addresses the impacts of flooding at the most local level. Little attention is given to watershed issues. The NFIP is coordinated through the state but no effort is made to address participating communities on a watershed basis. The CRS offers incentives for watershed planning but little guidance. It is also voluntary. The NFIP, CRS, and remapping efforts do not incorporate expected climate changes or changes in land cover. The NFIP needs to become more watershed- and future-planning based. More specific recommendations are given in Appendix A.
- The USACE addresses flooding in numerous ways ranging from offering assistance with flood response efforts to the construction of massive public works projects. Section 404 of the Clean Water Act establishes a relationship between watershed storage and wetlands and rivers. The USACE and the U.S. Environmental Protection Agency manage this program jointly. Section 404 reviews, assesses, and comments on proposals by others, similar to Environmental Impact Statements. This program could provide the forum through which to manage river/floodplain/watershed interdependencies.
- NRCS offers a suite of programs. They are incentive-based and voluntary. A flood-prone community may benefit greatly from the increased participation of farmers in the Conservation Reserve Enhancement Program or the Conservation Reserve Program.
- The Council on Environmental Quality (CEQ) supervises activities involved with NEPA including guidance for Environmental Impact Statements. As part of the office of the president of the United States, it is very political. The NEPA process requires an initial scoping

meeting to offer early notice of projects. In practice, though, actions are often far along before such meetings are held. NEPA is procedural. Many states have adopted NEPA regulations for state actions, and many of these do have substantive requirements, but they, too, are reactive. CEQ could develop more proactive strategies.

- FEMA administers recovery and mitigation programs. These programs are risk-based or address direct damages resulting from a Federal Disaster Declaration. In the main, FEMA repairs damaged, publically owned infrastructure and gives a little money to homeowners. It also provides mitigation funds to homeowners and public entities, which offer greater long-term benefits than reoccurring disasters cost. FEMA provides money for planning. All states and numerous cities have produced Hazard Mitigation Plans. These plans must be updated every five years. New requirements could be added to address the needs of climate change and increases in impermeable ground cover, as well as the need to address flooding and floodplains within the context of watersheds.

- The Small Business Administration provides loans to homeowners and businesses to cover disaster-related expenses that are not insured or assisted by FEMA funds.

- The U.S. Forest Service manages 193 million acres of public lands in national forests and grassland. The effect of forest practices on downstream interests has not been a driving force. However, potential changes in the species makeup of our forests and grasslands may change retention and detention rates in many watersheds. The USFS must change its emphasis to include those affected by changes in the resource.

- Executive Orders issued by the president have the effect of law and provide clarification to executive branch agencies. President Jimmy Carter issued EO 11988. This order, like NEPA, provides a process that all federal instrumentalities must go through if building in or affecting a floodplain. However, the order goes further than NEPA in preventing federal agencies from building in or adversely affecting floodplains unless there is no practicable alternative. The order could support planning efforts but has not been enforced this way.

- Basin commissions can offer opportunities to address changes in climate and land cover as they have in supporting agriculture, power generation, and transportation. The U.S. government has created basin commissions, the most recognized being the Tennessee Valley Authority and the Bonneville Power Administration. These can be models for the creation of other such authorities. States also have the authority to create similar special districts and basin commissions, and have done so.

Case Study: Flooding of I-5 in Washington

During the first week of December 2007, record-setting precipitation dumped on the Pacific Northwest. Torrents from water-gouged hillsides broke levees and overtopped dikes as floods reached record highs. One man was swept away in the deluge. The Interstate freeway (I-5) in Lewis County, Washington, was closed for three days and traffic was rerouted an extra 400 miles. The Department of Transportation estimated the cost of the I-5 closure exceeded $4 million per day. This was not the first such flood. In 1986, 1990, 1991, 1996, 2000, and again in 2009 the Chehalis River drowned the communities' lives in muddy water. I-5 was closed and traffic rerouted in 1990, 1996, 2007, and 2009.

Flood-prone communities located between Portland, Oregon, and Seattle, Washington, along I-5 have recently experienced growth. Although economic expansion continues to be important, much of the remaining developable land is in the floodplain.

Since 1996, Lewis County has granted more than 100 permits for new development in the floodplain. The cities of Centralia and Chehalis added more development. This included big-box stores, restaurants, strip malls, a railroad line extension, parking lots for a church, a coal-unloading facility, a new natural-gas pipeline, a mine expansion, barns, homes, carports, and shops. Many counties, including neighboring Thurston, have either banned or seriously diminished development in the floodplain; Lewis County and the two cities have not. According to a National Marine Fisheries Service statement, Chehalis has 9 percent of its Urban Growth Area in mapped floodplain, while Centralia has 21 percent.

A 1999 study analyzed flood peaks of the Newaukum River, a tributary of the Chehalis River, from 1942 to 1996. It showed that during that time, the Newaukum flood peak flows increased 21 percent, storm volumes increased 10 percent, and the same amount of precipitation increased flood peaks up to 22 percent in the later years. Precipitation was held constant for the study, so the increases are attributable to development within the watershed. These results are not directly applicable to the larger Chehalis basin, but show how quickly flooding patterns can change in that area.

The demand for water in the Chehalis watershed has steadily increased each year. This increased demand for limited water resources has made the water rights–allocations process complex and controversial. On many days each year, required base flows are not met. Predicted change in runoff from climate change will further challenge the system.

There is very high hydraulic continuity within the Chehalis watershed, meaning water easily flows between groundwater and rivers. Rainwater can recharge surficial aquifers. This provides the opportunity to store winter water for summer downslope needs, but includes the risk of surface pollutants contaminating groundwater. Water quality is degraded in parts of the Chehalis River and several of its tributaries during the summer months. Increased water use will further degrade water quality and fisheries habitat.

Chinook, chum, coho, and steelhead salmon and cutthroat and resident rainbow trout reside within the Chehalis River watershed. The Washington Department of Wildlife summarized the conditions of wild stocks of anadromous fish in the Chehalis as generally healthy with some depressed stocks in the tributaries. The levels of future populations will depend on stressors from contaminated water and low flows. Factors that will affect water quality and quantity include logging, mining, dams, diversions, obstructions, and commercial and residential development near streams.

1. What Values or Assets Do You Want to Protect or Enhance?

There were five principal stakeholders and five sets of values driving discussions of risk reduction.

- Department of Transportation wants I-5 open.
- Developers want to expand the built environment.
- Residents want to live flood-free.
- Farmers need water during the summer.
- Fish interests want cleaner water and adequate levels in the stream at the right times.

2. What Are the Apparent Risks or Opportunities for Enhancement?

- For the Washington Department of Transportation, the risk was the loss in dollars and political capital. Floodwaters over I-5 stop traffic and there are few alternative routes. The cost in dollars to the trucking industry, time to passenger cars, and increased greenhouse gases was exorbitant.

- Developers needed to determine an acceptable level of risk. For land developers, capabilities included realizing the potentials of the existing market, using insurance to manage risks, and lobbying for a level local governmental regulatory playing field.
- For residents the hazard was clearly flood damage. There may be opportunities for FEMA buyouts, but there were few replacement homes in their price range. They have few capabilities and were the least resilient of the set of stakeholders mentioned.
- Increased usage and climate change may reduce the amount of instream water in the summer. This can affect both irrigation and fish. Changes in water quantity and quality can significantly reduce fish runs.

3. What Is the Range of Risk-Reduction or Opportunity-Enhancement Strategies Available?

In 2002, the U.S. Army Corps of Engineers led a collection of agencies in developing the Centralia Flood Damage Reduction Project. Alternatives from that report included:

- No action.
- Modify an existing dam, the Skookumchuck, to allow for additional flood storage within one tributary facility.
- Overbank excavation to reroute flood flows.
- Upgrade the existing levee system with an emphasis on setback levees.
- Install flow restrictors to store water upstream of the restrictor.
- Promote flood-proofing structures, evacuation plans, and removing structures from the floodplain.
- Create an organization to implement an integrated strategy that includes parts of each of these alternatives.

These action items were driven by the first objective of the project team: reduce flood hazards in the project area to the maximum extent practical. Other objectives:

- Avoid increasing flood risks downstream from the project area.
- Avoid decreasing any existing low flow benefits provided by Skookumchuck Dam.
- Be cost-effective for both construction and maintenance.
- Avoid adverse impacts to the aquatic environment.
- Minimize and compensate for unavoidable adverse impacts to the aquatic environment.

- Incorporate appropriate fish and wildlife habitat creation, enhancement, and restoration measures.
- Comply with all federal, state, and local regulations, including environmental regulations.

Each alternative was evaluated on the feasibility of incorporating appropriate fish and wildlife habitat measures. Not discussed were forest practices, future expected changes in climate, or advantages of using upstream flow restriction structures and upstream storage to enhance winter flows.

No action was implemented after the 2002 study. However, following the December 2007 storm, the project was reactivated, and the original alternatives provided a point of embarkation. Work is ongoing and the flood of 2009 should ensure that interest continues in finding solutions.

4. How Well Does Each Strategy Reduce the Risk or Enhance the Resource?

The Washington State Department of Transportation wants a flood-free I-5 corridor. This value could be achieved through a combination of alternatives:

- Detaining water within the higher watershed.
- Storing water over the floodplain.
- Keeping discharges off the freeway.
- Raising the roadbed.

Land developers need development opportunities. They would benefit from solutions made at public expense that would create land, not take it (levees, dams, and dredging). Nonstructural measures might help them as well.

Existing residents would not gain from nonstructural approaches enforced through police powers such as zoning because these would benefit the owners of new construction. Nonstructural measures such as elevating homes above flood levels would be acceptable if outside money, such as through FEMA, would be available. Homeowners, however, might prefer dams, levees, and dredging because of their familiarity.

Fish and farms need clean water and summer flows. For them, upstream storage would be the best option. This could be accomplished with changes in forest harvest practices, upstream flow restrictions, and off-channel floodplain storage.

Including climate change and future development would alter the benefit-cost estimates dramatically. The Chehalis is heavily influenced by changes in rain intensity and changes in land cover and land use. Considering climate change and increased floodplain development would strengthen the argument for upper watershed solutions.

5. What Other Risks or Benefits Does Each Strategy Introduce?

The 2002 project team recommended a three-pronged strategy.

- Setback levees.
- Skookumchuck Dam modification.
- Nonstructural approaches.

The addition of setback levees would protect existing residential and commercial structures, I-5, and other transportation infrastructure from flooding. In 2002, large areas of the floodplain were not developed, making this alternative practical. This strategy is becoming increasingly difficult as local government allows additional floodplain development.

With small modifications, the Skookumchuck Dam could provide a couple thousand additional acre-feet of storage. This would help all stakeholders. Like all dams, however, it would concentrate water within the reservoir and not offer rewatering benefits thoughout the forested and agricultural watershed.

Nonstructural measures depend on the cooperation of the communities involved; the community has resisted such measures in the past. Counties that border Lewis have adopted the nonstructural measures proposed by the 2002 project team. Lewis County has not.

Setting back the levees places extra land on the inside of the levee system. This increases aquifer recharge, may decrease summer drought conditions for downstream properties, absorbs river energy, and provides additional habitat and possible amenities for the community. These setback levees, however, would reduce land available for development. Since 2002, some of that land has become home to megastructures.

6. Are the Costs Imposed by Each Strategy Too High?

Doing nothing will simply push costs into the future and increase them. The characteristics of the watershed are changing.

The upstream flow restriction structures and upstream storage alternatives were not considered by the 2002 project team but should have been. Their exclusion was in part because such measures would have expanded upstream floodplains. In light of expected increases in needs for summer storage and for increased winter flooding, this alternative should be considered. Those in the lower watershed could compensate owners of upland storage areas for their losses.

7. The Decision

The reinstituted project team now has a new mandate and expanded political capital.

This is not an issue that will have a single, ultimate project to solve all problems. A wide range of strategies will be needed to deal with this hazard. A project to specifically address keeping the freeway open is in the design phase, though funding has not yet been secured. Because the amount of development in the basin is increasing and the amount of available water is decreasing, cost-effective strategies must consider future conditions.

Chapter 9

Choosing the Best Strategy

No one strategy is perfect for all situations at all times. Because rivers serve so many functions, few stretches of water are devoted exclusively to irrigation, drinking water, recreation, waterfront development, and so on. Managing streams must include the flexibility needed to accommodate future changes. Several decision-making tools can be used to guide discussions. Tool sheets are included in this chapter to help you quantify the effect of your proposed strategies on the flood characteristics, the consequences of your strategies to your site and others, and your capabilities. Both current and future costs should be considered before making a decision.

Choosing a Strategy

It is an unfortunate truth that not everyone can win in every situation. When there are winners and losers, the best policy is to have the winners compensate the losers. For example, if an urban homeowner lives in an area with such frequent flooding that she must move, then either the city buys her out, she keeps collecting on flood insurance claims until FEMA changes its policy, or she gets tired of living with flood damage. In this case, one strategy would be to store water in the higher watershed in order to manage the discharge. The winners, urbanites, could tax themselves to buy flood easements, build detention facilities, or purchase harvesting rights from upstream stakeholders.

Every situation involves multiple options. Some reduce risk faster. Others are cheaper. Still others produce a greater number of benefits. How can you decide which options are best for your specific situation?

Managing the Change

We suggest assessing risks and opportunities by analyzing the flood, the effects of the flood (both beneficial and harmful), and your capabilities. You can assess your alternative strategies by looking at each of these items to see if addressing the flood will reduce risk or provide opportunities over the long term.

We have tried to make the case that we live in a constant state of change. Rivers continually erode, meander, and transfer sediment. Climates change. Old buildings are replaced and new uses replace old ones. Open land is converted into cities. The resiliency of natural and man-made systems depends on our ability to understand and build flexibility into both.

Natural physical and biological processes help us reduce risks and enhance benefits. These processes give us capabilities. All strategies must be driven by natural processes if they are to succeed over the long term. Recognition of these processes cannot be an afterthought.

Strategies can be grouped into three broad categories:

- Those that focus on how well your approaches alter the flooding (the change in the environment). Do they lower the frequency, reduce the affected area, reduce the severity, or give you more time to react? This set of strategies can include prevention or structural strategies.
- Those that focus on the effects of flooding. These include the non-structural approaches of retreat, accommodation, and protection.
- Those that focus on capabilities. Capabilities refer to approaches and tools that are affordable, timed appropriately, and may require available political power.

Examples include:

- A levee is built and focuses river energy through a smaller corridor (altering the location of change). You must then armor the bank.
- A levee is built to 100-year protection and reduces minor to moderate flooding (altering the frequency and severity of change). You must then pay for continued operation and maintenance.
- A river is straightened, shortening its length and increasing its ability to mobilize larger debris (altering the severity and timing of the change). You must then build sediment retention structures.
- A large reservoir is built in an effort to capture lost snow and ice storage, and in the process alter downstream flow regimes and trap sedi-

ment in the reservoir (altering the location of the change, creating secondary issues, requiring outside funding programs, and using political capital). Eventually, the reservoir has to be abandoned or dredged.

- We kill wolves, then elk eat riparian vegetation. Beavers leave, vegetation no longer anchors the riverbank, discharges increase, and more sediment is released causing the scouring of riverside homes (altering the severity and frequency of change, and altering the political climate to allow reintroduction of wolves and beavers). Homeowners must relocate homes or build canal protection structures.
- Some homeowners living along a river build bulkheads to prevent channel migration and to maintain grassy lawns. Native, more resilient, riparian vegetation and native animal populations disappear (altering the location of change). Ground cover must then be artificially maintained at expense to the owner and other homeowners must armor riverbanks (requiring the use of the capability of money).

Understanding and taking advantage of and managing natural processes can have beneficial results.

- Planting willows located along riverbanks can slow discharges, reducing erosion and destructive forces downstream. Even when bent from extreme velocities, willows can redirect erosive water energy.
- Wetlands, and natural or man-made depressions located in the higher watershed, can reduce downslope erosion and flooding.
- The creation of low-head dams and side-drain soil-retention structures can reduce downstream flooding.
- Lengthening streams, often by re-creating old meanders, can store water and reduce velocities, benefiting downstream stakeholders.
- Placing instream obstructions such as rock or large woody debris (LWD) can reduce downstream energy, benefiting downstream interests and creating instream habitat.
- Planting riparian areas and using soft river edges to connect habitat can take advantage of natural processes that spread and store water, recharge aquifers, and create vegetative friction.

At this point you must ask yourself whether implementing each of your strategies will create more problems. As we have said, no one strategy is perfect for all situations at all times. We may strive for strategies that result in no adverse effects but there are often winners and losers. Sometimes there are no winners, especially over the long term.

Decision-Making Tools

There are several commonly used decision-making tools that can be applied. Three are the scientific method, the National Environmental Policy Act (NEPA) approach, and benefit-cost analysis. Each addresses a different decision. The scientific method helps identify an approach that will work. NEPA identifies and informs others of the effects associated with an approach, and a benefit-cost analysis can help determine if the project is affordable over the long term.

Scientific Method

The scientific method is a decision-making approach that uses available data to help determine whether a project will actually achieve its intended objective.

The applicable steps in analyzing potential strategies are:

- Observe the situation. *You've done this by seeing that floods present a problem.*
- Define the question. *You've done this by identifying the value or asset you are trying to protect.*
- Gather information and resources. *You've done this by gathering information on all the potential strategies available to you.*
- Form a hypothesis. *Do this now by saying, "The appropriate strategy is . . ."*
- Collect data. *Take the strategy you chose and see what would happen if it were applied to your exact situation. Your starting point might be how it has been used in other cases, but you must draw up a comprehensive list of possible effects in your case.*
- Analyze data. *Take your list of possible effects and see which are likely, which are beneficial, what the costs are, and so on. Also, look at what would happen if nothing were done.*
- Interpret data and draw conclusions that serve as a starting point for new hypotheses. *Make a judgment on whether you want to keep that strategy as a possibility, and start the process over with the next strategy on your list. Keep repeating this process until you have comprehensive information and analysis on all your possible strategies. Only then do you decide on what your ultimate strategy will be.*

Following the scientific method to decide whether to elevate your home on posts and piers, you would:

- Look at other homes and see what happened to them during a flood.
- See why other homes in the same area were not flooded.
- Look at outcomes—why did some houses flood and others did not? Some houses don't flood because of location, others because they're elevated. It's important to look at the right set of comparables.
- Note all the disparities between the flooded and nonflooded houses. One by one, toss out the ones that don't apply to your situation.
- Do your data show that elevating a home made a difference? If not, discard this option. If it did make a difference, you should then research whether it's economically feasible and what financing might be available.
- Repeat this process for other available options (upstream storage, and so on).
- After investigating all options, make a decision.

National Environmental Policy Act (NEPA)

NEPA requires that the effects resulting from a project have been determined and explained. It proposes a process where a group of stakeholders meets to discuss a problem and proposed action. All alternatives are examined, including taking no action. If the action will have significant effects, an Environmental Impact Statement is required. If not, a less intensive environmental assessment is required. When identified, adverse effects must be mitigated if possible.

This includes:

- The environmental effects of the proposed action.
- Unavoidable adverse environmental effects.
- Alternatives, including no action.
- The relationship between short-term uses of the environment and maintenance of long-term ecological productivity.
- Irreversible and irretrievable commitments of resources.
- Secondary or cumulative effects of implementing the proposed action.

NEPA offers a process, not substantive requirements. If the process results in something potentially apocalyptic, NEPA has no mechanism to stop it. The strength of the process comes with disclosure, which triggers applicable laws so that adversely affected stakeholders are notified. We have incorporated these six NEPA concepts into the six questions we suggest you ask and answer before undertaking flood management programs.

Following NEPA for a strategy such as elevating your home on posts and piers, you would:

- Invite all those who have a stake in the project to a meeting to help define the scope of the project and all possible consequences.
- Research and verify all potential effects, as well as the laws and regulations that may apply.
- If there are significant effects identified, conduct an in-depth analysis of other options, including doing nothing.
- If there are irreversible effects, look into the possibility of mitigating these.
- If mitigation is not possible, identify another alternative.

Benefit–Cost Analysis

Benefit-cost analysis is another decision-aiding approach, and one that is required of most large government agencies. The process assesses whether the project benefits outweigh the costs.

This process is rigorous and usually demands that you quantify benefits and costs in dollars. Intangibles, such as a view, must be translated into monetary terms, which may be a source of considerable disagreement. FEMA has free benefit-cost software that they require in assessing various grant applications; this software can be used in other contexts as well.

In a benefit-cost approach, for a proposal to elevate your home on posts and piers to avoid high water, you would look at a potential future event (like a ten-year flood).

- Add up all the benefits, using dollars:
  —Anticipated damage that will be avoided if the home is elevated.
  —All related benefits such as having a better view, not having to worry about being flooded, increased market value, and so on.
  —All indirect benefits such as having reduced the flood effects to downstream stakeholders because you incorporated low-impact development (LID) methods in your design.
- Add up all of the costs, in dollars:
  —Cost of elevating your home.
  —All related costs such as living with your Aunt Louise while your home is being elevated.
  —All indirect costs such as disturbance to the riparian corridor while the home is being elevated.

- Annualize the dollar amounts and adjust for inflation.
  - —Anticipated damage doesn't happen every year so you'll need to divide your damage estimate by the annual flood frequency. If your building is in the ten-year floodplain, you may expect $100,000 in damage every ten years, or $100,000 divided by ten years—$10,000 every year.
  - —To annualize the construction costs, divide them either by how long you plan on living in the house or by the useful life of the project. Banks use thirty years when figuring the life of a home.
  - —Future project costs and money values will be different, so you'll have to adjust for inflation.
- Add up all of the annualized benefits and divide them by all of the annualized costs.
- If the result is greater than 1, the project is said to have a positive benefit to cost ratio and should be considered for implementation.

For example:

- Benefit of project: You expect to avoid $30,000 in damage every ten years, so your annualized benefits would be $3,000 per year ($30,000 divided by ten years).
- Cost of project: It costs $50,000 to elevate your house and you plan on living in the house for twenty years, so your annualized cost would be $2,500 ($50,000 divided by twenty years).

In this example your benefit-cost ratio would be 1.2 ($3,000/$2,500) and this is a good indication that you should proceed. However, this example does not include a host of indirect benefits, such as living without danger from floods, not living with Aunt Louise during floods, or the days taken off to clean up the property after floods. It also does not include some indirect costs such as living with Aunt Louise while the house is being elevated or potential environmental degradation should a flood come during construction and wash those materials into the water.

Tool Sheets

Included are four tool sheets and a summary sheet that can be used to assess your alternative strategies. These integrate the scientific method, analysis of effects, and benefit-cost analysis. They can help you evaluate your alternative strategies' appropriateness, effects on others, and whether you have the capabilities needed to actually implement your preferred strategy.

Using the Tool Sheets

These tool sheets assume you know all of the issues involved, can actually assign values to them and weigh each similarly, and that we have appropriately devised the tool sheets. It is rare to have information that complete. These tools are simply guides—an attempt to be a little objective with a very difficult subject.

You can use them in several ways. You can fill out each tool sheet as we have, using a scale of 2 for positive, 1 for neutral, and 0 for negative outcomes. Add up the results and divide by the appropriate number. A result of 0 to 0.99 suggests the situation would be made worse by the strategy. A total of exactly 1 would indicate no net change. A result greater than 1 would indicate a positive solution and the highest score among several strategies would be the best alternative. Remember, though, that these numbers are only rough approximations for complex problems. Ultimately, the best decision is usually made by a cross section of stakeholders.

You may feel more comfortable not using numbers but merely checking the items that are significant. Another approach would be to check just the items that are negative and determine whether these can be mitigated.

The first two tool sheets address whether or not your strategies will actually achieve their intended goal. The third addresses whether you have adversely affected others and the fourth assesses whether or not you have the capabilities to implement your preferred strategy. A fifth sheet has been added to help you summarize your results.

To demonstrate these tool sheets we have created a case study involving a small, flood-prone community. Homes have held their value because of their proximity to a regional market and to summer river recreation. The community is flooded quite often by a river draining about 1,000 square miles with flows originating in a low mountain range. A block of fifty flood-prone homes is at risk, as are adequate summer flows for recreation. The homeowners are somewhat concerned about summer flows because they promote the recreation that serves as the economic base of the community, but this is a secondary concern.

A few homeowners have banded together and are considering three strategies. These are not the only possible strategies, but a simplified example for illustration purposes.

- Strategy 1: Construct many small retention facilities throughout the watershed—a structural, prevention type of strategy similar to Buck Hollow River, Oregon (see the case study in Chapter 6).
- Strategy 2: Elevate existing structures above flood levels expected in the future—a nonstructural, accommodation type of strategy similar to Snoqualmie, Washington (see the case study in Chapter 2).
- Strategy 3: Do nothing—except buy flood insurance.

Assessing Flood Characteristics

Table 9.1 is a tool to help evaluate whether or not altering the characteristics of flood by applying natural biologic and physical processes helps you achieve your intended objective.

- The first column lists the values/assets this community wants to protect or enhance; the risks/opportunities that face them; and their possible strategies.
- Columns 2 through 5 rate the effect of each strategy in terms of how the strategy alters the hazard—the flood characteristics.
- In column 6, we tallied the scores for each strategy and divided by 4.
- The last column is used for comments.

In the above example, Strategy 1 is construction of retention ponds. It would reduce the frequency, extent, and severity of future floods, receiving a score of 1.6. It is an alternative worth seriously exploring. Strategy 2 is elevating a home, which does not affect the characteristics of the flood; this receives a neutral score of 1.0. Strategy 3 is to do nothing. Because of increasing development and projected changes in climate, flooding is likely to get worse. The final score of 0.0 reinforces this conclusion.

Assessing the Effects of Flooding

The assessment tool in Table 9.2 will help evaluate whether your strategy will be effective by addressing the consequences of the flood to natural, built, and societal systems.

It is often easier to reduce the risk by focusing on the consequences of flooding by retreating (moving buildings off the floodplain) or accommodating (elevating buildings above the floodplain or constructing a built environment that is not adversely affected by the change).

Remember, a flood is not just high water. Flooding can bring debris, velocity, or sediment; cause erosion or redistribute nutrients; create

Table 9.1
How the Example Strategies Affect Flood Characteristics

	Flood Characteristics									
	Will altering the characteristics of the flood by using natural biological and physical processes help you preserve or enhance your value/asset?									
	Frequency		Location		Severity		Timing		Scoring	Comments
	A		B		C		D		2=positive 1.5=moderately positive 1=neutral 0.5=moderately negative 0=negative $(A+B+C+D)/4$ Totals greater than 1 suggest positive change.	• Are similar strategies used in similar conditions? • What are the long-term effects to natural resources and processes? • What are long-term maintenance costs? • Did you consider all secondary hazards?
Value/asset: Reduce flood damage and increase storage **Risk/opportunity:** Floods damage homes but can also recharge aquifers and supplement summer flows	Comment	Score	Comment	Score	Comment	Score	Comment	Score		
Strategy 1: Construct many small retention facilities throughout the watershed to store water and reduce discharge	Doesn't reduce flows for 100-year floods, but does reduce smaller flood flows (10- to 25-year events)	1.5	Will reduce area flooded in smaller floods; may not change extent of 100- to 500-year flooding	1.5	Major benefits will be in smaller, more frequent floods	1.5	Warning time will increase	2.0	1.6	Strategy could also include preserving mature vegetation in upper watershed and riparian zones or increasing channel length and allowing for the collection of LWD

Strategy 2: Elevate structures above expected future flood levels	No effect on flood frequency	1.0	No change in amount of area flooded	1.0	Will not affect flood heights or secondary hazards like erosion	1.0	No effect	1.0	Elevating without fill, without increasing the footprint, and using other LID practices would not change streamflow or flood patterns	1.0
Strategy 3: Do nothing	Changing climate and land cover will increase future flooding	0.0	Similar frequency floods will cover increasingly more land	0.0	More energy from more frequent and higher floods will increase damage	0.0	Rivers will become more flashy, with decreased warning time	0.0	Will result in increased flooding, damage, and adverse secondary effects from stress to plants and animals; diminished summer flows may threaten recreation economy	0.0

Table 9.2
How the Example Strategies Affect Flooding

	Effects of Flooding						
	Will your strategy be effective in addressing the effects of flooding using natural biological and physical processes?					Scoring	Comments
	Systems					2=positive 1.5=moderately positive 1=neutral 0.5=moderately negative 0=negative (A+B+C)/3 Totals greater than 1 suggest positive change.	• Are similar strategies used in similar conditions? • What are the long-term effects to natural processes? • What are long-term maintenance costs? • Did you consider all secondary hazards?
	Natural		Built		Societal		
	Comment	Score	Comment	Score	Comment	Score	
Value/asset: Reduce flood damage and increase storage **Risk/opportunity:** Floods damage homes but can also recharge aquifers and supplement summer flows							
Strategy 1: Construct many small retention facilities throughout the watershed to store water and reduce discharge	Retention will reduce lower frequency events and distribute water throughout watershed, re-creating some snow cover benefits	1.5	Facilities may be difficult to maintain. Will increase summer flows for downstream use	1.5	Reducing floods will improve the lives of floodplain residents. Increases in summer flows will provide benefits, including agriculture and recreation	1.5	Re-creating beaver habitat and introducing beavers merits consideration

Strategy								
Strategy 2: Elevate structures above expected future flood levels	Detention on land that can safely flood provides benefits, but elevation may not remove potential hazardous debris (cars, gasoline cans, etc.) from property	1.5	Disruption and damage from floods will be minimized, but residents will have to relocate during floods	2.0	Floods will still disrupt community, but for more limited periods	1.5	1.7	To be successful, the elevation will have to consider future climate change and development of watershed
Strategy 3: Do nothing	Over long term, doing nothing will force retreat as homeowners must move because of frequent flooding, will benefit natural system, but not homeowners	0.5	Flooding will disrupt community for extensive periods, with severe and widespread damage	0.0	Flood victims and social, political, and economic services that serve them will be repeatedly affected	0.0	0.2	Systems supporting flood reconstruction such as insurance will become increasingly strained

habitat complexity; provide rich seedbeds; and recharge groundwater. Your alternatives should be assessed against all of these consequences, both negative and positive.

- The first column lists the values/assets this community wants to protect or enhance; the risks/opportunities that face them; and their possible strategies.
- Columns 2 through 4 rate the effect of each strategy in terms of how the strategy affects various systems.
- In column 5, we tallied the scores for each strategy and divided by 3.
- The last column is used for comments.

In the above example, Strategy 1 is building retention ponds. It would benefit natural, built, and societal systems, giving a score of 1.5. Strategy 2 is elevating structures above flood level, a better protection for the homes at risk in this example, giving a score of 1.7. Both strategies are worth pursuing. Strategy 3 is to do nothing. Because of increasing development and projected changes in climate, the effects of flooding are likely to get worse. The final score of 0.2 reinforces this conclusion.

Assessing the Effects on Others

It is relatively easily to determine your risks and opportunities and develop an approach that addresses your values and assets. It is more difficult to determine the values or assets of upstream or downstream stakeholders or forecast degradation of natural values that may not be apparent for many years. However, we have a legal (and perhaps ethical) responsibility not to cause adverse impacts to others and we often rely on government to reduce this liability through regulation.

Table 9.3 will help you evaluate whether your strategy will adversely affect other stakeholders or natural processes.

- The first column lists the values/assets this community wants to protect or enhance; the risks/opportunities that face them; and their possible strategies.
- Columns 2 through 4 rate each strategy's effect to upstream, floodplain, and downstream stakeholders.
- In column 5, we tallied the scores for each strategy and divided by 3.
- In column 6, we noted whether there is a possibility of mitigating adverse effects.
- The last column is used for comments.

In the above example, Strategy 1 is constructing retention ponds, which would help upstream, floodplain, and downstream stakeholders, receiving a score of 1.7. Strategy 2 is elevating structures above flood level, which would help floodplain and downstream stakeholders, receiving a score of 1.5. Strategy 3, doing nothing, would benefit downstream communities and receives a score of 0.5.

Strategies 1 and 2 help different stakeholders, but do not damage any. If resources allow, they might be complementary actions.

Assessing Capabilities

Table 9.4 is a tool to help evaluate whether or not you have the capability to implement and maintain your objective.

With the first three tool sheets, you looked at whether or not the strategy would produce your required results. The ratings for various strategies would be the same no matter who you are. However, not everyone has the same resources. The results can be substantially different depending on who you are or what you represent. Here you will look at whether or not you have the capabilities to implement your potential strategies.

- The first column lists the values/assets this community wants to protect or enhance; the risks/opportunities that face them; and their possible strategies.
- Columns 2 through 4 rate the community's capability to implement each strategy in terms of money, power, and timing.
- In column 5, we tallied the scores for each strategy and divided by 3.
- In column 6, we noted whether there is a possibility of increasing their capabilities.

In our example, Strategy 1 is constructing retention ponds. The people filling out the tool sheet own only a few pieces of land. They have not yet convinced the city's population as a whole that action is needed. This group does not have access to the political power or money to build more retention ponds, as shown by a score of 0.3. If this is the most appropriate strategy for the watershed, this small group will have to get other stakeholders involved to implement it. Strategy 2 is elevating homes. After a flood, the timing will be right because of increased mitigation resources, leaving a score of 2.0. Assuming a homeowner can wait until a post-flood period, this is an option he can afford. Strategy 3, do nothing, receives a score of 0.0 because it does not require new capabilities. In

Table 9.3
How the Example Strategies Affect Others

Effects on Others (NAI assessment)

What are the positive and negative effects of your strategies on others?

Value/asset:	Upstream		Floodplain		Downstream		Scoring	Mitigation	Comments
	A		B		C		2=positive 1.5=moderately positive 1=neutral 0.5=moderately negative 0=negative (A+B+C)/3 Totals greater than 1 suggest positive change.	Is there a possibility of mitigating adverse effects on others? How?	Have you considered the: • Environmental sensitivity; • Scale, and • Long-term effects of the project?
Reduce flood damage and increase storage									
Risk/opportunity: Floods damage homes but can also recharge aquifers and supplement summer flows									
	Comment	*Score*	*Comment*	*Score*	*Comment*	*Score*	*Score*		
Strategy 1: Construct many small retention facilities throughout the watershed to store water and reduce discharge	Can increase storage, increase soil moisture, support irrigation and available drinking water	2.0	Can reduce frequency of flooding, with limited impact on large (100-year) floods	1.5	Can reduce frequency of flooding, with limited impact on large (100-year) floods	1.5	1.7	Allow LWD in channel to benefit downstream interests, while purchasing development rights to upland areas, which might then flood	Larger, less frequent floods could be accommodated through a blending of Strategies 1 and 2

Strategy 2: Elevate structures above expected future flood levels	No effect	1.0	Beneficial to residents of elevated structures	2.0	Allowing site to flood will provide storage and reduce discharges from frequent flooding	1.5	1.5	Could result in release of hazardous materials, which could require policing; a warning should be included in strategy	This strategy could address a wide range of floods
Strategy 3: Do nothing	Climate change can reduce snowpack, increasing floods and decreasing summer flows	0.0	Climate change can reduce snowpack, increasing floods and decreasing summer flows	0.0	Flooding of area results in water storage, offering some benefit to downstream communities	1.5	0.5	Climate change can change watershed biology, which can increase flooding; need mitigation measures for fire, etc.	Flood frequency may lead to community abandoning floodplain and river would reclaim it (a possible benefit overall), at expense of homeowners

Table 9.4
How the Example Strategies Are Affected by Stakeholder Capabilities

	Capabilities						Scoring	Increasing your capabilities:
	Do you have, or can you get, the capabilities to implement and maintain your strategy?							
Value/asset:	Money		Political power		Timing		2=positive 1.5=moderately positive 1=neutral 0.5=moderately negative 0=negative (A+B+C)/3 Totals greater than 1 suggest positive change.	What incentives or partnerships can you make to increase your capabilities?
Reduce flood damage and increase storage	A		B		C			
Risk/opportunity: Floods damage homes but can also recharge aquifers and supplement summer flows								
	Comment	Score	Comment	Score	Comment	Score		
Strategy 1: Construct many small retention facilities throughout the watershed to store water and reduce discharge	Strategy includes land outside jurisdiction and requires money to purchase land or development rights.	0.0	Needed land outside community control, but could ally with upper watershed interests (farmers, timber, etc.)	0.5	Community should exploit a significant flooding event to increase political power and funding	0.5	0.3	This strategy requires involving others, so homeowners could organize with others and increase their political strength, or tax themselves to fund the upland retention structures

Strategy 2: Elevate structures above expected future flood levels	Homeowners may be able to use their own resources, or use FEMA's suite of assistance programs	2.0	Many flood reduction programs already exist and homeowners could organize to use them in their community	2.0	If homeowners were recently flooded, they could exploit increased interest and disaster assistance programs	2.0	Homeowners have control over own property and few other resources (except for money) are needed	2.0
Strategy 3: Do nothing	Increasing damage through time will increase recovery costs	0.0	As flood damage increases, community could organize for help, or give up and fracture	0.0	As flood damage increases, the most resilient households will leave, fracturing the community	0.0	Significant damage from floods and lack of summer flows will continue to increase until market factors force a retreat from the floodplain	0.0

some cases, however, doing nothing will actually cost money (in future flood damage) or will cause the homeowner to miss out on other resources that are available only for a short time.

If a city or county were filling out this tool sheet, the scores could be considerably different. A local, state, or federal government unit might have considerable capabilities and decide to build retention ponds. This could, in fact, be just one part of an integrated plan to deal with problems in a watershed.

Total Cost of Each Strategy

Three strategies were analyzed on the four tool sheets. Add the sheets together and divide by 4. Table 9.5 shows the result.

Strategy 1, building retention ponds, scored a total of 1.3.
Strategy 2, elevating a home, scored a total of 1.6.
Strategy 3, doing nothing, scored a total of 0.2.

Strategy 2 had the highest score, largely because our stakeholders' capabilities (those of a few homeowners) were limited. If our stakeholders had more resources, Strategy 1 might be preferable as it provides more benefits. A larger group of stakeholders to work on these issues could include farmers, fishers, timber interests, recreation businesses and users, and other communities along the river. The best case might be a long-term plan to implement both Strategies 1 and 2. Strategy 3, the do-nothing option, will allow the watershed to degrade and floods to get worse.

As a final question, we are asking you to determine if the costs of your strategies are too high. The cost can be:

- *Monetary.* The long-term costs of maintaining levees or riprap can be unexpectedly high.
- *Economic.* Neighborhoods or business districts that are located along rivers can lose economic value if the river or floodplain is degraded.
- *Shifted onto others.* Does your strategy solve your problem but create new problems for upstream or downstream users?
- *Degradation of natural processes.* Wetlands are a natural system that can filter pollutants. If they are destroyed, that filtration must be paid for in water treatment processes.
- *Fragility.* A barrier to keep out floodwaters may not hold up if the flooding gets worse.

- *Insufficient planning.* Many times, the best location for flood management will be the upper watershed, but there are few watersheds that are located entirely within one jurisdiction. To get around that, communities sometimes force a project within their own boundaries, even though it only partially solves the problem at a higher than necessary cost.
- *Aesthetic.* Some people value the presence of nature, particularly in an urban environment. Levees can reduce a vibrant riverine biota to a monoculture of grasses and rodents and also block the sight of running water.

Many other costs must be balanced when choosing a strategy. If it were an easy process, there would be many more success stories. As we learn more about the interaction of the river with plants, animals, and humans alongside it, the "best" choice will probably change through time. The same is true as cultural values change. River views, for example, are more important now than they were in the nineteenth century when rivers were used as sewers.

Resiliency is another critical component of successful strategies to combat flood damage. There is a relationship between the scale of the change and the recovery time of natural systems. A new structure may have a limited effect on the watershed as a whole, but the effects may last for a very long time. Similarly, a new structure may have a significant consequence at the site itself, but that consequence may not last very long.

An easy way to think of your strategy is in terms of losses prevented or benefits gained. This could be in terms of time, money, and quality.

- *Time*: How often might this loss occur or benefit be realized?
- *Money*: Will the cost of the flood damage outweigh the cost of the strategy over the long term?
- *Quality*: How much risk prevention and opportunity enhancement can you afford, both emotionally and in dollars?

If the benefits do not justify the costs, you have to redesign your strategy.

Make the Choice

There is one additional question to our list of six: Which strategy do you choose? It's important to keep this as a final, separate step. In general,

Table 9.5
Summary of Tool Sheet Scoring

	Summary sheet				
	Flood characteristics	*Effects of flooding*	*Effects on others*	*Capabilities*	*Total*	*Comments*
Value/asset: Reduce flood damage and increase storage	A	B	C	D	(A+B+C+D)/4	
Risk/opportunity: Floods damage homes but can also recharge aquifers and supplement summer flows						
Strategy 1: Construct many small retention facilities throughout the watershed to store water and reduce discharge	1.6	1.5	1.7	0.3	1.3	This strategy tempers the adverse effects of climate change, and represents a greater benefit to downstream properties. However, our example community has limited capabilities (lacks the needed funding and political clout to influence decisions outside their community)

Strategy 2: Elevate structures above expected future flood levels	1.0	1.7	1.5	2.0	1.6	Strategy 2 receives a higher rating primarily because the example community has greater capabilities to elevate their own homes than to increase retention/detention on land owned by others
Strategy 3: Do nothing	0.0	0.2	0.5	0.0	0.2	A do-nothing alternative is not beneficial for the community in question, but may be the best alternative for downstream interests over the long term as our subject community abandons their increasingly flood-prone site and the river reclaims the floodplain

you should keep your options for strategies alive until you complete the analysis by going through all six questions. Some strategies will ultimately be rejected because of their cost in terms of money, political power, time, or ineffectiveness.

The choice you make will be one that you, society, and the environment will have to live with for a long time.

Case Study: Tulsa, Oklahoma

The Tulsa, Oklahoma, experience demonstrates how we as a nation have addressed flood risks, how our values have progressed, and how our approaches to changing flood risks need to evolve further.

Tulsa has battled floods for most of its history. Due to the dedicated efforts of its citizens and government, Tulsa is now generally recognized as having the best floodplain management program in the nation. It is one of only two communities with the highest CRS level of 2 as of this writing, which brings a 40 percent flood insurance premium reduction to its residents.

Tulsa evolved with flooding. The city sits on the bank of the Arkansas River and lies in a weather zone defined by violent storms. A few significant events include (all dollars in 2000 values):

- 1923—Arkansas River flooded Tulsa's waterworks causing $5 million in damage and leaving 4,000 homeless.
- 1943—USACE built levees to provide flood protection.
- 1960s and 1970s—Floods struck every two to four years.
- 1970—Mother's Day flood in Tulsa caused $725,000 in damage on rapidly developing Mingo and Joe creeks.
- 1976—Memorial Day flood killed three and damaged more than 3,000 buildings; total damage of $120 million.
- 1984—Memorial Day flooding was the worst in the city's history killing 14, injuring 288, and damaging or destroying nearly 7,000 buildings for total damage of $300 million.

At the start of the twentieth century, the city responded to flooding with Tulsa's first land-use plan, which envisioned upland boulevards and housing. In the lowlands, such as Mingo Creek east of town, generous parks and recreational trails were planned. This advanced the value of setting flood-prone land aside for storage and public enjoyment, and living with—even enjoying the benefits of—floodplains and flooding. It was a true multi-objective approach to floodplain management. As a result, the city Waterworks Plant moved to higher ground and a band of creek bottomland (Bird Creek) became one of the nation's largest city parks. The farsighted preservation of Tulsa's 2,800-acre Mohawk Park was destined to save the city innumerable future flood losses.

In response to the Great Mississippi River Flood of 1927, Congress passed the Lower Mississippi Flood Control Act in 1928 authorizing the U.S. Army Corps of Engineers to construct dams and levees to control flooding. This shepherded the extensive levee system built along the Arkansas River, designed to protect the Tulsa oil refineries.

The post–World War II building boom caused housing to fan out onto the floodplains to the south and east. Land that had periodically flooded with little harm was now awash in wave after wave of urban flooding. By the late 1950s flooding of newly developed subdivisions along the Arkansas River and its tributaries spurred calls for additional flood control.

Tulsa enjoyed another boom in the 1960s, when the city's population grew by 25 percent. Tulsa's rapid growth required pastures and meadows to be piped and paved as new buildings continued to spill into the lowlands of the creeks and streams that etch the area. In 1964, the USACE completed Keystone Dam 15 miles upstream from Tulsa. The rapidly urbanizing Mingo watershed was annexed to the city in 1966.

Between 1960 and 1970, the city flooded every two to four years. Confidence in relying solely on structural approaches waned in Tulsa. Then, on Memorial Day 1976, a 3-hour, 10-inch deluge resulted in floods that killed three. Citizens demanded action. Newly elected city commissioners responded. They enacted a floodplain building moratorium; hired the city's first full-time hydrologist; developed comprehensive floodplain management policies, regulations, and drainage criteria; enacted stormwater detention regulations for new developments; instituted a fledgling alert and warning system; and began master drainage planning for major creeks.

A new phase of stormwater management began with President Jimmy Carter's 1978 Water Policy Initiative. It recognized the need to place nonstructural techniques on a par with flood-control structures and to preserve the natural values of floodplains and wetlands.

To curb continuing losses, in the early 1980s the federal government developed the Federal Interagency Hazard Mitigation process. Within days after disasters, federal teams were dispatched to identify hazard mitigation opportunities—ways to make the response to each disaster reduce the scope of the next one.

The 1984 Memorial Day flood was the worst in Tulsa's history. Fourteen people died. The newly elected mayor and street commissioner had been in office for only nineteen days, but both knew the issues well. In the darkest hours of the city's worst disaster, they pledged to reduce the likelihood that such a disaster would be repeated. Before daylight, and with the help of FEMA, a Flood Hazard Mitigation Team was formed, and within days a new approach to Tulsa flood response and recovery was born.

The community focused on correcting the effects of flooding, including removing, raising, or flood-proofing the most vulnerable buildings, as well as setting aside flood-prone land for storage and public enjoyment.

Climate change is another hazard they must face. The Oklahoma Climatological Survey predicted for their state:

- The warm season becomes longer and arrives earlier.
- The cool season warms and shortens, which leads to a longer frost-free period and growing season.
- Earlier maturation of winter wheat and orchard crops leaves them more vulnerable to late freeze events.
- Year-round evaporation increases from the ground and transpiration from green vegetation.
- Drought frequency and severity increases, especially during summer.
- Drier and warmer conditions will increase the risk of wildfires.
- Rain-free periods will lengthen, but individual rainfall events will become more intense.
- More runoff and flash flooding will occur.

1. What Values or Assets Do You Want to Protect or Enhance?

Tulsa wants to continue to live with rivers while reducing flood damage and maintaining summer flows.

2. What Are the Apparent Risks or Opportunities for Enhancement?

Like other communities, Tulsa is facing new risks because of increased changes in land cover and climate.

Tulsa's population continues to grow, which will result in additional impermeable surfaces and runoff as well as increased water needs for residents. More flash floods can be expected.

3. What Is the Range of Risk-Reduction or Opportunity-Enhancement Strategies Available?

Dams and levees have been tried but were periodically overwhelmed. Because the most advantageous locations have already been developed, further large structures have an even lower chance of protecting the city.

However, Tulsa has a number of successful strategies that can provide the basis of future actions. In the 1980s a package of multi-objective approaches was implemented that reduced the hazard and its effects, and increased the community's capabilities. The flood hazard was reduced through the inclusion of:

- More than a quarter of the floodplain as open space, providing riverside detention areas to store water.
- Water detention structures, including dams.
- Nonstructural programs to replace and retrofit the built environment.

Implementation was incremental. Some homes were bought by the city with stormwater revenues. After floods, damaged homes were bought with the help of FEMA disaster relief funds. (Soon after the 1984 flood, almost 300 homes were purchased. Another 1,000 homes were subsequently purchased.) Response plans were developed and implemented. This process was repeated many times.

Other actions included:

- Establishing a stormwater utility.
- Setting up a watershed-wide floodplain management program.
- Approving millions of dollars for flood-control projects that are now complete.
- Developing and enforcing strong building codes, including the requirement of a 2-foot safety factor (freeboard) in floodplain construction.
- Community outreach to advise residents of flood hazards and offer mitigation solutions and technical assistance.

More capabilities are needed to combat future changes. Several should be researched.

To manage the flood characteristics:

- Purchase additional floodplain properties for detention to store floodwaters.
- Require LID to provide retention and spread water over as much of the watershed as practical.
- Use biological processes to store water and provide friction to dampen overland flow.
- Identify annually rechargeable aquifers and build detention opportunities within the aquifer recharge basin.
- Eliminate the floodway concept in appropriate areas where upriver storage and increased floodplain width is possible.

To reduce the effects:

- Design built environments to accommodate overland flow.
- Make greater use of home elevation and overland flow corridors.
- Institute water conservation programs.
- Require low-impact development to reduce damage to buildings.

4. How Well Does Each Strategy Reduce the Risk or Enhance the Resource?

None of these strategies are new, or even new to Tulsa. They just need to be expanded to address increased precipitation and increased discharges. As for drought, the extent of the expected change may be new to Tulsa, but tools to deal with it are not new to the West. Water must be conserved and stored. The city may have to look at flooding differently. Quickly flushing water downstream may be less important than buying, either in whole or just development rights, upstream property for water storage.

5. What Other Risks or Benefits Does Each Strategy Introduce?

Managing changes in discharge and climate will:

- Remove additional properties from the tax rolls.
- Inconvenience some residents who have to accommodate extreme surface flows.

- Extract a cost from property owners by requiring them to do things like build wet gardens, provide onsite storage, and adhere to other LID practices.
- Require a debate if political pressure is brought to build big structures that concentrate water in isolated areas and further stress natural biological processes.

In the process of reducing the effects there will be property owners who wait to be flooded before relocating structures above or outside of the floodplain.

6. Are the Costs Imposed by Each Strategy Too High?

The most cost-beneficial way to address complex issues is with an integrated plan of several strategies. Tulsa officials will have to judge the cost of strategies they formulate, taking into account both short- and long-term expenses. The cost of action will also have to be balanced against the cost of doing nothing and allowing problems to grow. Continued development and future climate change will exacerbate existing issues.

7. The Decision

Problems like increasing development and climate change are gradual, and major effects may not be apparent for years. Changing conditions mean that Tulsa—like other cities, counties, state, and federal agencies—must either increase their flood-control efforts or live with increased threats of more flood damage. Ironically, they may also have to cope with the inverse problem of low summer flows. However, Tulsa officials and community leaders have developed a community participatory infrastructure—a foundation that should be able to provide leadership to address these variations of old problems.

Chapter 10

What Next?

Floods cause damage but also create opportunities. We can preserve and even enhance our riverine values by taking advantage of rivers' natural processes. Both development patterns and future climate change can alter the amount of water, flow patterns, water quality, and storage potential of our watersheds. As water flows off the land and into our rivers because of the loss of snow, ice, vegetative storage and friction, wetlands, and permeable depressions, less water is stored for future use. Using all the information presented in this book, you can address problems and enhance your assets today and in the future—whatever it may bring.

River Values

Rivers are our lifelines. We cannot live without them. They replenish the soil with sediments and nutrients transported downstream. They recharge aquifers from which we take drinking or irrigation water. They provide a natural, complex environment that allows a variety of plants and animals to coexist with us.

Rivers also flood and cause damage. It is the nature of streams to overflow their banks periodically. On the other hand, we need rivers to act predictably and stay out of our cities. These two things do not have to be mutually exclusive. Rather than fighting a river's natural behavior, we can work with it to reduce the destruction and death caused by flooding.

In accommodating these expected changes, there are a few simple principles to keep in mind:

- Accept that rivers will do what rivers do.
- Realize that rivers are necessary to transport water, energy, sediment, nutrients, and nurture life.

- When trying to reduce the risk of flooding or enhance the benefits of rivers, look at the entire watershed. The surface water, groundwater, climate, geology, flora, fauna, and human uses of the land should all be taken into consideration.
- In general, the most effective and least expensive strategies are those that work with natural processes and focus on nonstructural approaches. These tend to have the fewest unintended secondary effects.
- Strategies that focus on adjusting the severity, frequency, location, or timing of the flood have the potential to create long-term problems. These strategies often rely on large structural measures such as dams, channelizing, or armoring channels, which can require expensive maintenance and result in extensive adverse effects over a very long time.
- The more that natural physical and biological processes are integrated, the less environmental degradation will occur. The greater the degradation, the more costly the project is over the long term.

Flood mitigation projects must be analyzed in the context of the entire watershed. Flooding may appear to be a problem only on the floodplain, but the cause of the problem may be land-use practices in the upper watershed. Some strategies, like adding fill to raise homes above the water level, help individuals but cause new problems upstream or downstream.

Scale is important when looking at flooding problems. The scale of the problem and its solution should be consistent. A problem that affects a few homeowners should not be solved by changing the dynamics of the entire watershed. Likewise, resolving and paying for a watershed-wide problem should not be borne by the few closest to the stream channel. For example, as a whole, it is less expensive and less intrusive to raise homes above the level of flooding than to build a dam or levee to protect a neighborhood. However, for the individual homeowner, remodeling can be very expensive and may require subsidies.

One of the future unknowns is our climate. Climate models predict a range of changes, and there is no clear answer as to what the climate will be in 50 or 100 years. What we know for sure is that flowers are blooming earlier in the spring or staying open later in the fall and species of both plants and animals are adjusting their ranges to reflect the local changes that have already taken effect.

This is an important point when considering river management projects. Our traditional method of flood control is to get water through the system as quickly as possible. If models are correct and some places end up with less snowmelt available in the spring or different rainfall patterns through the year, our streamflow patterns will change. We may have too much water in the winter but not enough in the summer. There may be an increased need to even out flows, to store more water on the land, and not let it—or make it—rush to the sea.

The answer is not to build more dams or levees. The answer is to take the steps that may be initially more difficult, because they are not as familiar. We must change the way we approach flood control, to make working with nature, instead of fighting it, an honored value and accepted practice.

Taking Action

In an interview for an American Planning Association newsletter article, Tulsa community leader Ann Patton suggested that planners remember five Cs to understand the dynamics of creating an effective floodplain management program:

- *Crisis.* It is hard for a community to ignore the problem when it is in a crisis. Seize the opportunity.
- *Coalition.* Bring together citizens, experts, professional staff, and news media and continue to broaden with everything from developers to bicycle trail enthusiasts. Look for people with primary interests other than floodplain management.
- *Commitment.* "I can hardly overemphasize this," she says. "These issues take years, and in my experience, at least a generation."
- *Comprehensive.* Don't look at problems in a piecemeal manner. Everything is connected, and the entire watershed must be treated as a complete natural system.
- *Combinations.* Marry structural and nonstructural solutions, finding the optimal balance for each specific problem.

Ultimately, the strategies used for enhancing streams are the same whether you're a citizen, government body, or part of a group interested in a river. There is power in numbers and if you want to make changes beyond the property you own, you will need to join with others. As Ann Riley writes in her book *Restoring Streams in Cities: A Guide for Planners, Policymakers, and Citizens*, you need to know about a lot of different fields: civil engineering, hydrology, geomorphology, environmental regulations, citizen involvement, and ecology, to name a few. A team of people will probably make the task easier and more enjoyable, and will develop a stronger (more efficient, less expensive, longer lasting) project.

In looking at possible flood management strategies, what other problems can be addressed, or what other riverine assets can be enhanced? Are people in your area worried about water for agriculture? Recreation? Drinking? Tourism? Some of the "hooks for community involvement" listed by Riley include:

- Control erosion and reduce flood damage.
- Enhance the neighborhood and re-create a sense of community.
- Attract tourism and revitalize the downtown economy.
- Preserve history and cultures.
- Increase fishing opportunities.
- Create jobs, job training, and educational opportunities.
- Create trails and greenways.
- Mitigate for land-use changes.
- Reclaim ecological values.
- Restore water quality.

Once you have an idea of what you want to happen, the team building begins. Which existing groups might care about the health of the river? These could include organizations as diverse as formalized watershed councils, Riverkeepers or other environmentally based groups, neighborhood groups, hiking or bicycling clubs, fishing or angling groups, schools (primary grades to college) that encourage their students to do community or environmental projects, garden or historical societies, and local or state government or special districts. Once you start asking people if they're interested, you may be surprised at the amount of feeling people have for their rivers.

At this early stage, you should also begin to educate yourself about your watershed. What trees, shrubs, plants, and animals existed in its natural state? What exists now? How many buildings are flooded because they are in the river's natural floodplain? How has development changed the flooding patterns of the river? How many dams and levees exist? Some of the places to go for this information are your local government planning or development department, state geological agencies, U.S. Environmental Protection Agency's Surf Your Watershed, Natural Resources Conservation Service Soil Surveys, and National Flood Insurance Program maps. You must understand the characteristics of your watershed before any meaningful project can be developed.

From that point on, the size and complexity of your project will determine the number of stakeholders to involve and the time it will take to complete. Whatever the size of the project you conceive, it will be important to work with your local government. If you can, interest locally elected officials or staff who can help you through the process of implementing some sort of environmental change. Typical activities to interest local residents and governments include having a stream cleanup day, coordinating a tree-planting volunteer project to reduce erosion, and convincing local media to do a story on the state of your stream. You will need patience and perseverance. Remember, the problems took years to develop; they cannot be undone in a moment.

Six Questions

The point of this book is not to give you an easy checklist of projects to choose from. We would if we could, but life is more complicated than that. Instead, what we have given you is background information on the natural processes you need to consider.

But we have also given you a list—that is, the six questions to ask before you decide on how to reduce the risk of flooding, or how to enhance riverine environments. Hopefully, at this point, the flow of the questions makes sense.

1. What values or assets do you want to protect or enhance?
   • What's important to you?
2. What are the apparent risks or opportunities for enhancement?
   • How is it being threatened or how can you improve it?
3. What is the range of risk-reduction or opportunity-enhancement strategies available?
   • What natural, engineered, or societal systems can be used to develop strategies to protect or enhance your values or assets?
4. How well does each strategy reduce the risk or enhance the resource?
   • Do all the possibilities actually protect you or enhance the resource over the long term?
5. What other risks or benefits does each strategy introduce?
   • Do your solutions bring up new problems or provide other benefits?
6. Are the costs imposed by each strategy too high?
   • Is your strategy worth the cost?
7. The decision is:
   • Do your homework and your analysis, but don't forget to use your

intuition. Remember that involving all stakeholders will result in more support for the strategy, and may give you ideas that you wouldn't have thought of.

And don't forget:
Enjoy the river.

Case Study: Rivergrove

People love living and working in Rivergrove. The jewel of the downtown core is the riverfront. On one bank is a wide swath of park, including an outdoor amphitheater, tree-lined paths, and markers showing the history of the city and its river. The other bank is protected by a harder edge, though the seawall was taken down. Now the concrete barrier is made of steps that allow access to the river but still protect the historic buildings. Warehouses from the past, when river commerce was the lifeline of the city, look familiar but have been converted to modern office buildings and residential condos.

Rivergrove's economy is strong, as its beauty; safe, cheap drinking water; and lack of flooding attract a skilled labor force that employers need. It is also a destination for tourists and recreational river users who bring additional money to the region.

The water that flows through town sparkles. In the upper watershed, forestland and best forestry practices keep the sediment load near its natural levels. Agricultural practices in the lower basin also take water quality into account. The river is clean enough for cities throughout the basin to take their water from it with minimal treatment. Fish runs are stable and plentiful after being almost nonexistent for years. Tributary streams that flow into the main river have enhanced wetlands, which also protect water quality even as they absorb and lower flood peaks.

Tiling, draining, and paving over wetlands almost destroyed them. Now they feature a complex web of plants and animals that aids the river and provides recreational and educational opportunities for Rivergrove residents and visitors. The school district spends a week each year teaching sixth-grade kids the importance of the riverine system to Rivergrove, starting with field trips to the wetland areas. Using the river as a focus, teachers use science and math to explore real-world problems, show how humans and the environment coexist, and ask kids to use the environment as an inspiration for art, writing, and music projects.

Even though the upper watershed is only a few miles from the urban area, it is full of wildlife. Native trees, grasses, and flowers host native animal species. Beaver have come back to the river, building their dams and small impoundments, creating habitat for a plethora of other animals. Raptors patrol the skies.

Swale enhancement and setting back levees have allowed the river to regain much of its original floodplain. This lessens flooding and allows water to pond and augment summer flows. In the lower watershed, many buildings were bought and razed to reduce the damage and risk from flooding. Land-use practices keep people from building in the floodplain so new construction is not at risk.

As the largest urban area in the watershed, Rivergrove took the lead in organizing a watershed-wide council made up of local governments, state and federal government agencies, urban and rural business interests, recreational users, and residents. The council is able to look at the watershed as a whole, exchange information among its constituent groups, and influence land-use and other policy decisions throughout the watershed. They were instrumental in getting one of the two dams on the river removed. The remaining dam is necessary for irrigation and summer flows, and serves as a point for water recreation. It is managed to mimic the natural pattern of streamflow with higher flows during the traditional months of flooding, even though some water is kept in the reservoir for low summer flows.

The watershed council is one way of ensuring this idyllic setting will last into the future, as is educating the children of the area. More important, however, as residents, farmers, ranchers, other business interests, visitors, recreationalists and others enjoy the amenities of the basin, they each have a stake in making it work. Rivergrove will continue to attract people who value it for generations to come.

Appendix A
National Flood
Insurance Program

The National Flood Insurance Program (NFIP) makes no attempt to control the river, but instead focuses on reducing the adverse impacts of water on the floodplain. Although it is a federal program, it is enforced through lending institutions and local governments. Some critics feel that NFIP encourages building in floodplains because it insures against flood losses. Suggestions for improvements to the program are given.

A Huge Program

The significance and the breadth of participation in the National Flood Insurance Program are enormous.

Insurance

- Average annual flood losses in the United States are currently estimated at $6 billion.
- Between 1978 and September 31, 2007, the NFIP paid $33.2 billion for flood insurance claims and related costs.
- The average claim paid for the years 1997 through 2006 was $46,168.
- Homeowner's insurance seldom covers flood damage.
- More than 5.5 million people currently hold flood insurance policies, in more than 20,300 communities across the United States.
- A $100,000 flood insurance premium on a house would be only about $33 a month.
- If you live in a FEMA-designated Special Flood Hazard Area (SFHA) or high-risk area and have a federally backed mortgage, your mortgage lender requires you to have flood insurance.

Mapping

- FEMA flood maps exist for more than 19,000 communities across the United States and its territories.
- These maps were the result of the Flood Disaster Protection Act of 1973 and many are more than thirty years old. The state of the flood maps is inadequate. FEMA is in the midst of map modernization, but in many places these maps do not fully reflect the true risk to life and property.

Regulations

- NFIP communities have adopted, and are enforcing, floodplain management ordinances to reduce future flood damage.
- Over one thousand communities participate in the Community Rating System (CRS), with two communities as of this writing qualified for insurance premium discounts of up to 40 percent.

A Brief History

Toward the end of the 1920s, the emphasis turned to construction of large structural solutions to address flood problems. Good flatland was increasingly hard to find and communities were moving into flood-prone areas. Our technological knowledge was increasing and giving us the confidence (often arrogance) to embark on very large public works projects.

By the 1960s, we began to realize that these big public works projects were very expensive to maintain and didn't work all that well in dynamic environments such as rivers and floodplains. Our flood-related damages were not decreasing. A new approach was needed. Gilbert White, sometimes called the father of floodplain management, developed one. He argued for focusing on reducing the effects of flooding.

When it was created in the late 1960s, the National Flood Insurance Program represented a major shift in policy, based largely on White's ideas. It moved away from the policy of trying to "fix" the river with large structural projects to one focusing on nonstructural solutions and providing incentives to reduce the effects of flooding.

The Basics of the NFIP

The NFIP has three major components.

- *Insurance*: Flood insurance is available in participating communities that require new and substantially damaged buildings to comply with a set of flood risk–reduction measures. Purchasing insurance is required if the federal government was involved in the construction or financing.
- *Regulation*: Risk-reduction regulations are required for new and substantially improved structures.
- *Mapping*: FEMA prepares floodplain maps that reflect existing conditions and regulates new and substantially improved structures located within this mapped area.

Insurance

Before the NFIP, the private insurance industry could not sell affordable flood insurance policies because only those at high risk would buy them.

In 1965, as a result of the devastating effects of Hurricane Betsy on the Gulf Coast, Congress passed the Southeast Hurricane Disaster Relief Act. The 1965 act was designed to provide financial relief to flood victims. It also authorized an insurance feasibility study entitled *Insurance and Other Programs for Financial Assistance to Flood Victims*. The study would help pave the way for the development of the NFIP.

Prior to 1968, the federal government attempted to control coastal and riverine flooding on a national scale through rechanneling, using dams and levees to restrict the flow of waters, and developing hydroelectric power and irrigation projects that had flood-control benefits. However, the increasing costs of these projects and high annual totals of flood-related damage influenced the government to explore the possibility of decreasing disaster relief payments through flood insurance.

The National Flood Insurance Act made federally subsidized flood insurance available to owners of improved real estate or mobile homes located in a floodplain if their community participated in the NFIP. Modeled loosely after homeowner's fire insurance, the NFIP was a voluntary program from 1968 until the adoption of the Flood Disaster Protection Act of 1973. But, as floods occurred, it became obvious that few were buying insurance—so it became mandatory.

As long as a community participates in the NFIP, flood insurance is available both within and outside of floodplains. Flood insurance is required if the federal government is involved in improved real estate (insuring loans through regulated lenders) and the land is located within a SFHA.

Insurance premiums differ for new and old construction. For the purposes of the NFIP, *new* refers to structures built after the Flood Insurance Rate Map (FIRM) went into effect and are referred to as post-FIRM construction. Post-FIRM structures pay rates based on real risk and these actuarially based rates can be very high. Rates for insurance on such structures can reach 25 percent of the value per year if built in violation of community NFIP regulations. Structures built before the FIRM became effective are called pre-FIRM and pay subsidized rates set by Congress that are not based on risk.

Insurers purchase contents and structural coverage from private property and casualty insurance companies. The federal government underwrites the risk. Policy premiums support the administration of the program.

The NFIP also offers insurance to cover the costs of meeting the building codes they require. This insurance is Increased Cost of Compliance (ICC) coverage. Currently ICC is limited to $30,000 and is restricted to four situations.

- *Elevation*—Raising a home or business to, or above, the flood elevation level adopted by the community.
- *Relocation*—Moving a home or business out of harm's way.
- *Demolition*—Tearing down and removing a flood-damaged building.
- *Flood-proofing*—Waterproofing, through a combination of adjustments or additions of features to the building, which reduces the potential for flood damage.

Regulation

For insurance to be available, communities must adopt and enforce floodplain management measures to regulate new construction. They must also ensure that substantial improvements to existing buildings within SFHAs are designed to eliminate or minimize future flood damage. The community's floodplain management ordinances must require:

- Building permits for all development within the SFHA.
- New and substantially improved residential buildings to be constructed with the lowest floor at or above the base flood elevation (BFE) if shown on the FIRM.
- Nonresidential buildings to be either elevated or flood-proofed to that elevation.
- The establishment of a regulatory floodway based on a maximum al-

lowable 1-foot rise in BFE and preventing improvements that would result in an additional rise.

Floodways

Floodways are important to the NFIP. They are linear areas set aside to convey the 100-year discharge without increasing the base flood elevation by more than 1 foot. Floodways almost always include the river channel and some overbank areas. The floodway designation assumes a built-out condition across the floodplain that results in a 1-foot rise in 100-year flood elevation. The encroachment on the river corridor that produces this rise defines the cross section of the floodway.

No structure is allowed in the floodway that would cause an increase in surface water elevation. The floodway designation benefits upriver interests by assuring them that structures will not be built downstream that have damming effects.

While floodways serve upstream development, they can have an adverse effect on downstream communities. Floodway requirements discourage in-channel storage beneficial to downstream interests and can harm river ecosystems. Energy-absorbing, friction-producing properties of in-channel structures, including natural features such as rocks and woody debris, are often prohibited in floodways.

Mapping

An SFHA is an area within a floodplain having a yearly 1 percent or greater chance of a flood. SFHAs are delineated on flood maps issued by FEMA for individual communities.

There are two basic SFHA maps. One is very general and often just a best guess at illustrating the 100-year floodplain. These Flood Hazard Boundary Maps (FHBMs) were first prepared by FEMA in the 1960s and 1970s and have largely been replaced by maps that are more detailed. However, FHBMs may still be the only maps available in rural areas with little expected development.

Flood Insurance Rate Maps (FIRMs) have largely replaced FHBMs. These maps are the product of rigorous analysis and contain information that is more detailed. FIRMs illustrate the 100-year floodplain boundary, the actual elevation the floodwaters are expected to reach during a 100-year flood, and often include the area expected to be inundated during a 500-year event. FIRMs in urban or urbanizing areas also illustrate the floodway. Maps can be amended by reanalyzing the hydrology and preparing a revised map. For single parcels, a letter from FEMA can be used to amend an existing map.

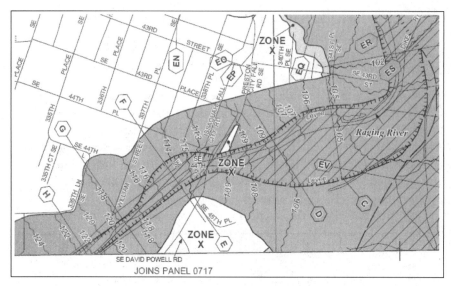

Figure A.1
This FIRM map shows a plan view of the valley, including the 500-year and 100-year floodplains and the floodway. (Source: FEMA)

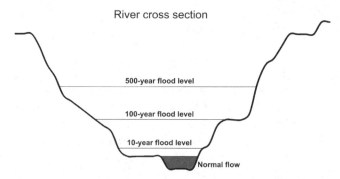

Figure A.2
Cross sections show expected river depth at various discharges. This is important information to include in flood management strategies.

Figure A.1 shows the width of the river and floodplains, including the regulatory floodway.

FIRMs also include river cross sections. Cross-section information is used to determine river water surface elevations at various discharges, as seen in Figure A.2. When cross-referenced with river profiles, they can provide more accurate base flood information than the map view.

A third river view is from the side or profile, as seen in Figure A.3. Profiles often show river restrictions and other hydraulic elements.

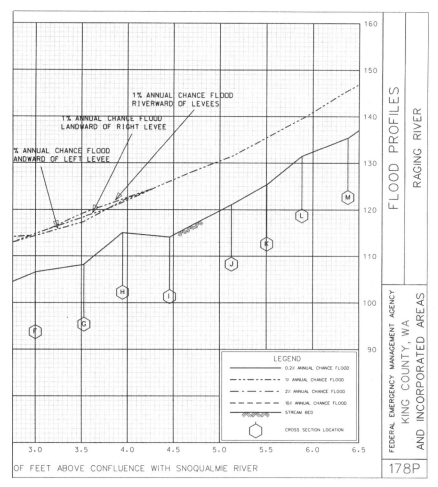

Figure A.3
A profile describes the river along a specific cross section, which is marked on the FIRM plan view. (Source: FEMA)

A hydrograph is a description of a river through time. Figure A.4 shows a stream with its normal flow and with a 100-year flood. Flows can be converted to river stages by using cross-section information.

Community Rating System

The Community Rating System (CRS) is an optional program that encourages floodplain management activities that exceed the minimum

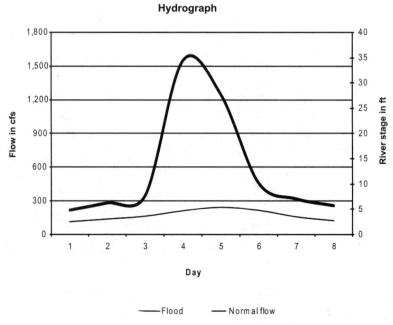

Hydrograph

Figure A.4
A hydrograph allows you to see how the amount of streamflow changes through time. It can be graphed for daily, seasonal, annual, or other periods.

NFIP requirements. Flood insurance premium rates are discounted up to 45 percent to reflect actions meeting the three goals of the CRS:

• Reduce flood losses.
• Facilitate accurate insurance ratings.
• Promote the awareness of flood insurance.

There are 1,028 communities receiving flood insurance premium discounts. They cover a full range of sizes, from small to large, and a broad mixture of flood risks including both coastal and riverine.

Potential Improvements for the CRS

The CRS could develop specific performance standards. Right now, the program is as much process as substance. Points are given for procedural elements such as making maps available or having a stormwater plan. More points translate to lower insurance rates.

With many communities facing rapid assaults on their existing FIRMs, specific performance standards are needed. Such specific, performance-based, substantive standards would have to incorporate standards and methods for determining future-conditions maps. They would also include the interventions that would contribute to wise management, such as low-impact development, watershed-based planning and management, wetlands preservation, the purchase of development rights, and so on.

Not Addressed by the NFIP

Approaches to reduce flooding losses that are neither supported nor discouraged by the NFIP include:

- Addressing downstream effects by discouraging the use of fill to elevate structures or encouraging water spreading to increase flood storage, dampen energy, and expand lateral biological continuity.
- Controlling lateral channel movement by installing friction devices to dampen in-channel and overbank energy.
- Addressing hazards that are not directly related to inundation: sediment transport, erosion, scour, water quality, and habitat issues.
- Addressing changes in flood effects over time, including changes resulting from increased development and its cumulative effects, land-use and land-cover changes, and climate change.
- New building technologies, including methods of flood-proofing residential structures, and designs for floating or temporarily floatable structures. Existing examples are found in The Netherlands and elsewhere.

Possible NFIP Improvements

- Structures are often rebuilt in places where adverse effects cannot be mitigated. Insurance allows flood-prone property to receive claim after claim. The NFIP should have a mandatory ceiling as to the number and total claims received.
- There is a need for watershed-wide management. Incentives, agreements, and additional legal tools that encourage watershed-wide management need to be developed. These could include mechanisms to allow communities to increase discharge retention through low-impact development (LID) and forest and agricultural practices.

- Regulation must include avenues to exploit natural processes that reduce risk.
- The NFIP should provide mandatory incentives to vacate the floodplain where mitigation solutions are not possible.
- The NFIP must discourage the use of fill to elevate structures and encourage the spreading of water over the land to increase flood storage, dampen energy, and expand lateral biological continuity.
- Regulations to manage development within the storage floodway need to be developed. This would include all development that did not reduce detention opportunities.
- Future-conditions floodplain maps should be required. The NFIP should require regulation to these future-conditions maps. Many communities currently manage their floodplains using such future-conditions maps, even though they are not required.
- FEMA has embarked on a billion-dollar nationwide mapping program called the Map Modernization Management Support (MMMS) program. A wide array of Flood Insurance Study (FIS) information is being digitized. The program is currently not funded to integrate updated hydrology but it is a platform with the potential of allowing low-cost map revisions as new hydrological information becomes available. At present the MMMS products reflect current conditions, but the MMMS program could make it easier to support the preparation of future-conditions maps.

Flexible Mapping

The MMMS program's ability to model expected flood events and depict them spatially makes it extremely flexible. MMMS can be applied to illustrate a variety of discharge scenarios. FEMA could use the MMMS program tools to generate two sets of maps: one reflecting the 100-year floodplain for insurance rating purposes and one reflecting future conditions that would be a far better basis for longer-term decision making about local land use, economic development, and other concerns.

- The future-conditions floodplain maps could be based on both natural conditions and human interventions. That is, communities could reduce the area designated as flood-prone on the future-conditions maps by applying NAI policies, LID practices, and other techniques to ensure that post-development discharges are less than or equal to predevelopment ones. In addition, communities could make use of natural depressions, wetlands, forests, and other existing watershed features to store water and preserve or even enhance their flood-reducing natural processes.

- The insurance maps would reflect current conditions so that flood risk and the appropriate insurance rates can be determined. With the MMMS tools, they could be updated regularly—perhaps every five years. Existing floodplain ordinances that meet NFIP standards would apply. Future-conditions maps could trigger a similar but more flexible set of NFIP regulations. These regulations could allow structures to be built below the future-conditions BFE if the structures could be easily retrofitted as projected flood levels rise.
- New maps, identifying watersheds along with detention and retention areas, and estimated rates of change are important.
- Secondary hazards such as debris, velocity, erosion, and meander belts need to be addressed.
- Opportunities for storage detention, vegetative retention and friction, and beneficial riparian corridors need to be identified.

Floodways

As discussed previously, floodways are useful in conveying water through a valued area while minimizing adverse upstream flooding impacts. They work if all that matters is getting water out of the way. But traditional floodway designations along an entire watercourse can have harmful effects on many of the attributes we value. Floodways:

- Are not designated with downstream impacts in mind—water reaches downstream communities faster with minimal reductions in discharge.
- Can reduce storage, lessening recharge opportunities.
- Often concentrate stream energy, which degrades channel stability, scours riparian areas, and leads to a change in the ecology of an area.
- May alter sediment transport mechanisms.

The quick-flush properties of floodways may be needed through developed areas but other floodplain values can be benefited by adopting a more comprehensive approach. Flexibility is currently possible.

Yet, removing the floodway without replacing this designation with other regulator guidance can also cause adverse effects. New built environments and accompanying disturbances can remove storage, cause increases in flood elevations, and cause water to back up onto land not otherwise flood-prone.

The NFIP and FEMA only map conveyance floodways. However, alternatives exist through *density fringe designations*, the ad hoc *storage capabilities of some floodways*, and even *zero-rise* designations. These help reduce at least two adverse floodway effects—lack of storage and the focusing of stream energy. Also, many communities have adopted *compensatory storage* regulations within floodplains including floodways.

Storage within Floodways

Floodways and their riparian corridors could provide the vegetative sandpaper to store water and dampen the energy harmful to streambeds and downstream communities. Traditional floodway calculations for river reaches with significant flows, low gradients, and broad floodplains often produce very wide floodways that convey floodwaters slowly, storing water in the process. Higher-frequency floods (those happening more often than a 100-year event) often pond within these broad, flat floodplains and store water, recharge aquifers, and drop sediment.

Zero-Rise Floodways

A zero-rise policy is more restrictive than a 1-foot-rise floodway, creating a wider floodway and a larger managed area. As mentioned above, for relatively flat reaches, floodways can provide opportunities for storage and accordingly downstream communities could adopt regulations that promote storage within such floodways.

However, a zero-rise floodway often provides additional protection to upstream development and may proportionately increase risks to downstream development.

Density Fringe Floodways

Traditional floodway designations can be unnecessarily restrictive for development within the meandering reaches or deltas of large rivers. Within such rivers, floodway designations are often proportionally large. The state of Washington does not allow residential buildings within designated floodways. They refer to density floodways as fringes. The engineering that defines them is identical to that for traditional floodways—only the regulatory requirements differ.

Regulation within a density fringe designation requires that:

- Structures be elevated above the base flood elevation.
- Structures be built such that the fill footprint does not exceed 2 percent of lot size within the density fringe.
- New developments (including fill) cumulatively cannot exceed 15 percent of a line drawn perpendicular to flow.

The density fringe concept is probably most beneficial where land-use constraints prevent structures from being clustered outside of a traditional floodway, where the floodplain is wide and where flows are significant.

Compensatory Storage

Of particular concern are commercial areas where big-box stores, parking areas, and service roads can easily remove all storage within the floodplain fringe.

Compensatory storage practices can offset some of the loss of flood storage capacity. When the developer puts new fill in the floodplain, it must be offset by excavating an additional floodable area to replace the lost flood storage area, preferably at hydrologically equivalent sites. In some cases, the developer must remove up to two times the amount of fill that is proposed to be placed in the fringe.

However, as we have learned in our attempts to mitigate impacts to or the relocation of wetlands, there are few hydrologically equivalent sites. Often the sites actually used are chosen for convenience and do not perform the same function of the ones being compensated for.

Case Study: Fife, Washington

The city of Fife, Washington, like many levee-protected communities involved in NFIP remapping efforts, is realizing that their levees no longer protect their floodplains. Some communities are threatened by such rapid changes that the FEMA flood maps cannot keep pace, insurance requirements do not provide adequate incentives, and regulatory requirements don't address emerging conditions.

Fife is a small community, largely commercial and industrial, just north of Tacoma. It lies at the mouth of the Puyallup River, which originates in the low reaches of the Cascade range. As is typical of snow- and rain-dominated drainage systems, the Puyallup experiences two flood cycles—one in the winter from seasonal rains, and one in the summer from snowmelt.

Since the end of the last glaciation, the Puyallup River has been a rain- and snow-dominated river system. However, with our changing climate, it is becoming a rain-dominated system. Much of the winter precipitation now stored as ice and snow in the glaciers that feed the Puyallup River could disappear in the coming years. The winter discharges (and base flood elevations) are projected to increase dramatically and the summer ones will diminish.

Estimates are that the winter streamflow peak will begin to exceed the summer peak near the year 2020. As the system becomes increasingly rain-dominated, summer drought will develop as the ugly stepchild of winter flooding. Drought changes the watershed. Wetlands dry up, soil-absorbing capacity diminishes, and stressed (or even dead) vegetated riparian corridors, forests, and wetlands do not anchor the soils, promote infiltration, or moderate flows in other ways. The result can be increased flooding.

Fife residents believe the levees have protected them from the river's flooding. FEMA contractors are now arguing that the levees are not up to standard and do not, in fact, offer adequate protection. In addition, existing FEMA flood maps do not reflect today's conditions and new maps are being prepared. The community believes the levees are strong and map revisions may not be needed. Residents are concerned that broader floodplain designations will discourage economic development.

1. What Values or Assets Do You Want to Protect or Enhance?

The community wants existing "protected" land to remain protected. They do not want to lose land to the floodplain and want a business environment that will not be interrupted with flooding.

2. What Are the Apparent Risks or Opportunities for Enhancement?

Fife's flood risks are changing rapidly. It lies in the crosshairs of two forces of change: climate and development. With increased development and predicted climate change, the Puyallup watershed is losing its ability to store water, but that prediction is not part of the discussion. FEMA currently does not consider future conditions when it produces flood maps for a community.

Contributing to this loss of storage, some cities in the Puyallup watershed are allowing subdivisions to be built with the most impermeable development possible—low building density, wide concrete streets, large houses that cover much of their lots, and massive stormwater systems that discharge directly into channelized river reaches.

3. What Is the Range of Risk-Reduction or Opportunity-Enhancement Strategies Available?

- Remove the floodplain designation.
- Maintain the levees and assume that FEMA concerns are unfounded.
- Fix the river with dredging, dams, and dikes. This has worked in the past and no additional land needs to be given back to the river.
- Remap the floodplain and rigorously enforce NFIP and CRS regulations. Outside the city of Fife, the county government sees the need for new maps because of river changes. It has assumed responsibility for the levee system. Pierce County has a very aggressive NFIP-driven program and has implemented many parts of the NAI initiative. The county has a CRS rating of 3 and should be upgraded to a 2 within the next review cycle.
- Take advantage of mitigation opportunities as they become available, such as replacing structures and infrastructure through regular maintenance or after disasters.
- Provide floodplain detention to reduce peak flows and help maintain current discharges.
- Prepare for expected changes by using the remapping opportunity to include expected changes from climate and increased land cover. A remapping effort could acknowledge future conditions in several ways, allowing for a long-term solution that might include:
  —Repairing the current levee system abutting intensely developed lands to perhaps seventy-five-year protection.
  —Reducing existing peak flows through detention along the floodplain corridor.
  —Managing floodplain creep through increased watershed-based, required retention.
  —Providing safe-fail mechanisms such as allowing for the flooding of the built environment.

4. How Well Does Each Strategy Reduce the Risk or Enhance the Resource?

- Removing the floodplain designation does not change the risk.
- The levee system has protected the city for years and there has not been a serious levee breach. This does not address future changes.
- Fix the river. This is an expensive choice, if it is based on increased levees or damming. Elements of this strategy, however, will probably be a part of any comprehensive strategy, especially through intensely developed areas.
- Remap the floodplain and rigorously enforce NFIP and CRS regulations. In some ways, this is an experimental strategy, but the city of Fife is located within Pierce County and is next to the Puyallup Reservation. Both implement NAI measures.
- Take advantage of mitigation opportunities. This strategy can be very effective so long as the interim damage is not to critical structures or interdependent systems that limit community resiliency and recovery.
- Provide floodplain detention. Many reaches along the Puyallup River currently have minimal development and could be set aside to temporarily detain water for flood storage. The levees along one several-mile reach have already been set back and storage opportunities were increased. If setback levees or increased floodplain storage strategies were pursued, tens of thousands of acre-feet of detention would be needed. This would be expensive to implement, but might be the most cost-efficient over the long run.
- Use the remapping opportunity to prepare for expected changes. This remapping effort could acknowledge future conditions in several ways, allowing for a long-term solution. To reduce future flood damage, a new set of regulations that target the watershed is needed. The SFHA can be maintained only if flood discharges remain the same. This would require increases in retention and detention to match increased discharges resulting from losses in snow cover and an increase in land cover.

5. What Other Risks or Benefits Does Each Strategy Introduce?

- Remove the floodplain designation. It would be in violation of the NFIP and flood insurance would no longer be available.
- Fix the river. The costs, including maintaining whatever structures are built, will exceed the cost of the lands being protected.

- Remap the floodplain and rigorously enforce NFIP and CRS regulations. Remapping the floodplain, upgrading the levee system, and rigorously enforcing the NFIP—even with higher self-imposed CRS initiated measures—will not address future expected changes.
- Take advantage of mitigation opportunities as they occur. The breakdown of the New Orleans levee system demonstrates the limitation of waiting. With many levee segments destroyed came the opportunity to repair them correctly, but at great cost. When the levees failed, entire communities were destroyed.
- Provide floodplain detention. Taking advantage of yet undeveloped possible detention sites along the river would be an asset for Fife currently and in the future. Fife should therefore pay for the purchase outright or for the development rights of such upland storage. The city may not agree or may not have the resident support to adopt appropriate and sufficient bond issues or other financial measures.
- Use the remapping opportunity to prepare for expected changes. This may be the most successful long-term solution, but there may be additional flood damage between now and the implementation of future programs.

6. Are the Costs Imposed by Each Strategy Too High?

Fife can achieve its desired value of having a residential and business environment that will not be interrupted with flooding. Over the long run, however, this will be prohibitively expensive if existing protected land remains protected and they are not willing to lose some developable land for flood storage. Costs involved in fixing the river and maintaining the fixes in a changing environment will be exorbitant.

Aggressively following the existing NFIP program may work in stretches but will fail with expected changes in weather and streamflow patterns. There may be valid arguments for following this strategy and repairing structures and infrastructure to known future conditions as they are destroyed. For residential properties, this approach is supported through insurance. The strategy fails for those dependent on uninsurable structures. Existing infrastructure is expensive to redesign and often drives a future risk-prone built environment.

A strategy taking advantage of mitigation opportunities as they arise after floods and as detention and retention purchase opportunities become available may work, but only if the community as a whole is willing to incur damage until those mitigation measures are taken. For resilient communities, this may be possible.

7. *The Decision*

This ongoing discussion has not yet been resolved. There are proponents for each strategy, from doing nothing to planning for future, larger floods. Because of the commitment of Pierce County to the NFIP, CRS, and NAI philosophies, they are likely to pursue remapping and seriously consider information about potential future hazards. Dealing with those hazards will probably require a combination of structural and nonstructural solutions.

Appendix B
Floodplain Designer's Tool Kit

Watersheds can be graphically represented by using a system of polygons, lines, and points. This gives you the flexibility to map current conditions and compare them to maps of what the watershed might look like after implementing a given strategy. This can help you see both intended and unintended consequences of particular strategies. Ultimately, this can help you make a decision about the most effective strategy.

This book addresses floodplains as products of their watershed, being dynamic and ever changing. We consider flooding not an inherent hazard but simply a disturbance or change in condition. We have presented the natural processes that produce change and offer ways, emphasizing these natural processes, to reduce adverse effects and enhance benefits. All of these concepts are presented in words.

Here we offer a way to sketch these processes. Knowledge of GIS helps but is not necessary. An inspiration for this approach comes from the work of noted urban designer Kevin Lynch, who defined a graphic notation system for the conceptualization and design of urban landscapes. Lynch's system, set forth in his 1965 book *The Image of the City*, consisted of point, area, and linear features such as nodes, landmarks, districts, pathways, and edges. Mapping an urban landscape in these terms greatly aids analysis of the components of urban landscape and their interrelationships.

Although the physical elements of a watershed differ from (as well as include) some of Lynch's urban components, a similar graphic notation system can be derived and applied in analogous ways. Such a notation system is presented below. Watershed processes and areas of concern can be expressed as polygons. Water transport processes and the methods to obstruct the transport of water, sediment, and nutrients can be expressed as lines. Assets such as structures and infrastructure can be expressed as points.

Its ultimate purpose is to help us think about change in watershed and how to reduce risk or take advantage of opportunities. Each watershed and flood plan presents unique conditions. This tool kit is a starting point, not a complete list of all variables to be considered.

Polygons

Water falls to the ground as rain, some of which is *retained* while some is *discharged* and flows into streams. Once in streams some water is *detained* upstream of restrictions, some flows over the banks onto *floodplains*, and some water provides the energy to reconfigure *active terraces*.

Large watersheds may begin in mountainous areas with heavy snow loads and may offer extensive retention that provides substantial flows throughout the year. Watersheds draining urban, sewered environments may offer very little retention, creating flash flood conditions and intermittent flows. Streams draining through arid, hard rock canyons may provide no water retention, resulting in high-energy water corridors that are dry or nearly dry most of the year.

Land uses change throughout the watershed. Steep slopes may be clear-cut, transitional land may be put into agriculture, and new development may take place on terraces.

Functions and Areas of Concern Represented by Polygons

These areas can be illustrated with polygons, as seen in Table B.1.

1. Functions
   - Retention areas
   - Discharge areas
   - Detention areas
   - Floodplains
   - Floodways
   - Active terraces
   - Other
2. Areas of concern
   (These are illustrated in Table B.2.)
   - Changes in physical condition, risk, or opportunity
   - Changes in biological condition
   - Changes in built environment

Lines

The vast majority of water falling to the ground as rain works its way downhill. This surface flow can be in *well-defined channels* or *undefined*

Table B.1
A Floodplain Designer's Tool Kit

Districts: Areas expressing similar water transport processes such as forested hills, flat meadows, towns, floodplains, wetlands.

a. Detention areas: artificial flow control structures used to contain flood-water for a limited period of time, thereby providing protection for areas downstream

b. Retention areas, high: surface storage areas having coefficients of runoff less than 30%. Woodlands or pastureland on sandy loam or mixed soil. Cultivated land that is flat and on sandy loam soil. Urban/suburban development including some parks, cemeteries, detention facilities, large lot subdivisions, low-impact development having less than 30% runoff.

c. Retention areas, medium: surface storage areas having coefficients of runoff greater than 30% but less than 50%. Woodlands on flat or rolling terrain or clay soils. Pastureland, rolling, or hilly terrain with mixed soils or clay soils. Cultivated land not on sandy loam soils. Urban/suburban residential development of medium lot size or smaller lots incorporating appropriate LID measures.

d. Discharge areas, medium: cities, towns, subdivisions, heavily cultivated areas having coefficients of runoff greater than 50% but less than 75%. Woodlands on hilly terrain with clay soils. Pastureland on rolling or hilly terrain mixed soil areas. Cultivated land, not on flat or rolling terrain, on sandy loam soils. Urban/suburban, all commercial, industrial, multi-use residential, and small-lot residential not incorporating LID measures.

e. Discharge areas, high: cities, towns, subdivisions, heavily cultivated areas having coefficients of runoff greater than 75%. Clear-cuts on hilly terrain with clay soils. Intensely urban/suburban with extensive sewer systems including commercial, industrial, multi-use residential, and small-lot residential not incorporating LID measures.

f. Floodplains: Designated by FEMA, 100-year floodplain.

g. Floodways: Designated by FEMA.

h. Active terraces

i. Other

---

channels. In some defined channels, flow is *intermittent.* In some cases, this flow becomes groundwater in *active interflow areas.*

These flow patterns help describe water transport processes. Arid areas may have very well-defined channels that have flows only a few months during the year. Karst areas may have extensive underground flows but few surface flows. Old-growth forests may have very few

Table B.2
Areas of Concern or Opportunity

A Floodplain Designer's Tool Kit

Places of concern, risk, or opportunity: Critical areas, areas in transition, areas transitioning from snow-dominated to rain-dominated, places of special significance, areas of opportunity, critically stressed biological communities.

a. Changes in physical condition, risk, or opportunity: erosion, tectonic uplift, stream reversing course.

b. Changes in biological condition: ground cover, stressed populations, succession to new flora or fauna, beetle infestation.

c. Changes in built environment: land-use changes, purchase/transfer of development rights.

expressions of surface flows. Sediment-rich valleys may have rivers and streams with extensive groundwater systems.

Functions, Structures, or Actions Represented by Lines

Water transport lines are shown in Table B.3.

1. Corridors—description and manipulation of paths that are parallel to flow
   Description of transport processes
   - Surface flows—undefined
   - Surface flows—intermittent
   - Surface flows—defined
   - Active interflow areas
   Manipulation of transport processes
   - Channel straightening
   - Channel lengthening (meanders)
   - Increasing discharge
   - Other
2. Obstructions to transport—often placed perpendicular to flow
   (These are illustrated in Table B.4.)
   - In-channel or near-channel control or energy dissipation features
   - Energy diversion, bank protection structures

Table B.3
Corridors

A Floodplain Designer's Tool Kit

Corridors: Transport paths.

    a. Undefined surface flows: overland flow, flows over riparian areas, poorly defined streams, streams with intermittent flows.

    b. Defined surface flows: typical streams, rivers, canals, etc.

    c. Channel straightening.

    d. Channel lengthening: meanders.

    e. Increasing discharge: quick flushes through built environments, increasing velocities.

    f. Very active interflow areas: hydraulic continuity with aquifer, streams flowing over deep glacial deposits, karst areas, lava flows.

    g. Connectivity between channel and floodplain.

    h. Other

- Obstructions to lateral flows
- Off-channel obstructions
- Existing obstructions to move
- Obstructions designed to control breaches
- Other

Points

Points can show items or places of value. Our example is a structure, but other possibilities for points include natural sites such as wetlands or ponds.

Table B.4
Obstructions

A Floodplain Designer's Tool Kit

Obstructions, edges, structures that restrict transport processes: Check dams, levees, groins, large woody debris (LWD). Structures could be set perpendicular to the channel thereby interrupting downstream flows or perpendicular to channel restricting lateral flow. Solid lines could represent harder obstructions such as concrete, wavy lines represent rocks, dotted wavy lines represent vegetative obstructions.

 a. In-channel or near-channel control or energy dissipation features or structures: detention reservoirs, pools, and riffles. (The bolder the line, the harder the structure—e.g., concrete and rock check dams as a solid line, while LWD, willows as dashed wavy lines.)

 b. Energy diversion, bank protection structures: concrete or riprap as a solid line, LWD or vegetation as hashed line.

 c. Obstructions to lateral flows: levees, bank protecting structures, berms.

 d. Off-channel obstructions: friction-generating structures including trees, shrubs, woody debris.

 e. Existing obstructions to move.

f. Obstructions designed to control breaches: contain channel, reduce erosive energy behind levee.

g. Other

Conceptually, to reduce the risk to a structure, you can move it out of harm's way, provide a protective shield for it, or build it in such a way that it will not be harmed. For a house, this means elevating it above the floodplain; building (or moving) it outside the floodplain; building a levee around the house; or designing the house so that getting wet would not harm it (flood-proofing).

The valued item may be a wetland that is at risk from urban development. Risk-reduction measures may include moving the wetland (this approach seldom works); providing a protective shield; or building filters and vegetative retrofits to accommodate the hazard.

Different situations demand different solutions. Some communities may be able to keep their floodplains free of structures; others may adopt an elevation approach. Still others with existing structures may encour-

Table B.5
Places of Value and Assets

A Floodplain Designer's Tool Kit

Places of value and assets.

a. Structures: built or natural feature.

b. Structures: elevated.

c. Structures: moved off floodplain.

d. Structures: protected.

e. Structures: flood-proofed.

f. Structures to raze.

g. Other

age owners to use flood-proofing methods. Points can be illustrated as in Table B.5.

1. Value
   • Structures
   • Other

2. Risk-reduction or opportunity-enhancement methods
   • Structures—elevate
   • Structures—move off floodplain
   • Structures—protect
   • Structures—flood-proof
   • Structures—raze
   • Other

Map Layers

Especially when using a GIS, gathering and integrating as many map layers as possible can significantly add to your understanding of the watershed. Table B.6 shows some types of layers that might be useful in watershed description and analysis.

Table B.6
Map Layers

A Floodplain Designer's Tool Kit

Map layers.

a. Contour.
b. Land use/land cover: Urban (residential, commercial, industrial), rural (woodland, pasture, cultivated).
c. Soils: clay, loam, sand.
d. Land ownership: private, public, parcel size, and location.
e. Groundwater: shallow and deeper rechargeable aquifers.
f. Floodplain and floodway: FEMA maps, community future conditions if available.
g. Other

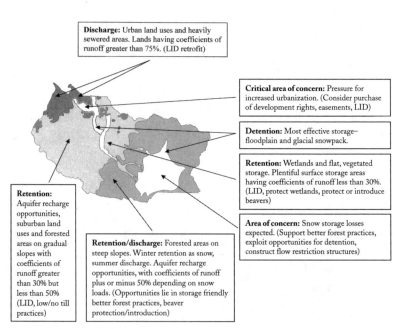

Discharge: Urban land uses and heavily sewered areas. Lands having coefficients of runoff greater than 75%. (LID retrofit)

Critical area of concern: Pressure for increased urbanization. (Consider purchase of development rights, easements, LID)

Detention: Most effective storage—floodplain and glacial snowpack.

Retention: Wetlands and flat, vegetated storage. Plentiful surface storage areas having coefficients of runoff less than 30%. (LID, protect wetlands, protect or introduce beavers)

Area of concern: Snow storage losses expected. (Support better forest practices, exploit opportunities for detention, construct flow restriction structures)

Retention: Aquifer recharge opportunities, suburban land uses and forested areas on gradual slopes with coefficients of runoff greater than 30% but less than 50% (LID, low/no till practices)

Retention/discharge: Forested areas on steep slopes. Winter retention as snow, summer discharge. Aquifer recharge opportunities, with coefficients of runoff plus or minus 50% depending on snow loads. (Opportunities lie in storage friendly better forest practices, beaver protection/introduction)

Figure B.1

Using the Tool Kit

The following examples illustrate how the floodplain designer's tool kit can describe large- and small-scale approaches and strategies. Each can be expressed as a GIS product or a hand-drawn sketch.

Three hypothetical examples are shown. Each applies the graphic symbols listed above. The first watershed-scale example illustrates how water runs off the watershed and into streams and rivers. The following two examples apply concepts presented in this book at a sub-basin scale and site scale. Both graphically explore the objectives of reducing flood damage while increasing the beneficial effects from storing water for future use at the surface and recharging shallow aquifers.

Watershed: Runoff Characteristics

Figure B.1 shows a watershed. Each box describes an area with specific characteristics. In parentheses are recommendations to improve the

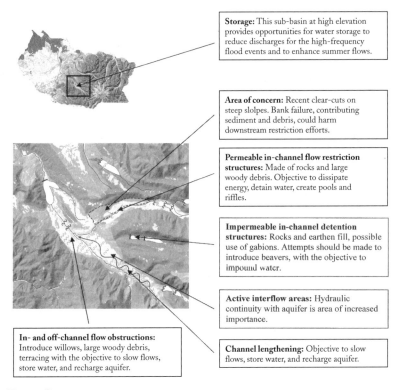

Storage: This sub-basin at high elevation provides opportunities for water storage to reduce discharges for the high-frequency flood events and to enhance summer flows.

Area of concern: Recent clear-cuts on steep slolpes. Bank failure, contributing sediment and debris, could harm downstream restriction efforts.

Permeable in-channel flow restriction structures: Made of rocks and large woody debris. Objective to dissipate energy, detain water, create pools and riffles.

Impermeable in-channel detention structures: Rocks and earthen fill, possible use of gabions. Attempts should be made to introduce beavers, with the objective to impound water.

Active interflow areas: Hydraulic continuity with aquifer is area of increased importance.

In- and off-channel flow obstructions: Introduce willows, large woody debris, terracing with the objective to slow flows, store water, and recharge aquifer.

Channel lengthening: Objective to slow flows, store water, and recharge aquifer.

Figure B.2

functioning of that area. (LID indicates low-impact development.) The first step in making this map was gathering GIS layers for elevation, 4-percent slope, wetlands and floodplains, land use/land cover, and urban areas. A similar process could be done by hand.

Sub-Basin: Storage

The entire watershed should be looked at for opportunities and solutions. The sub-basin in Figure B.2 has no urban development and little damage from flooding. However, it presents an opportunity to store water, which can significantly reduce flooding downstream including the major urban areas many miles away. Storage here can also be tapped in the summer when streamflows are low and there is demand for drinking and irrigation water.

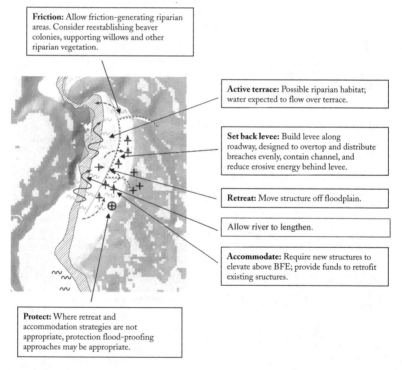

Friction: Allow friction-generating riparian areas. Consider reestablishing beaver colonies, supporting willows and other riparian vegetation.

Active terrace: Possible riparian habitat; water expected to flow over terrace.

Set back levee: Build levee along roadway, designed to overtop and distribute breaches evenly, contain channel, and reduce erosive energy behind levee.

Retreat: Move structure off floodplain.

Allow river to lengthen.

Accommodate: Require new structures to elevate above BFE; provide funds to retrofit existing sructures.

Protect: Where retreat and accommodation strategies are not appropriate, protection flood-proofing approaches may be appropriate.

Figure B.3

Site: Reducing Local Flood Risks

At the scale of an individual site, structures become important. When trying to reduce risks to homes, opportunities for increased upstream detention should be explored. This can reduce discharge downstream and reduce the danger to a wider area than acts like elevating specific homes. This example is not a suggested course of action for any specific community.

Further Reading

The concepts in this book are a tiny fraction of the information available about river systems and floodplain management. The authors drew on their various professional backgrounds and years of education, training, and practice. Here are a few resources we recommend for further reading or that we used for this book. Many others exist, through books, professional journals, government agencies, popular media, and the Internet.

Printed Material

Amoros, C. 2001. The concept of habitat diversity between and within ecosystems applied to river side-arm restoration. *Environmental Management* 28(6): 805–817.

Association of State Floodplain Managers. April 23, 2003, draft. *No Adverse Impact: A Toolkit for Common Sense Floodplain Management*. Madison, Wis.: Association of State Floodplain Managers.

Ball, J. A. 2004. *Impacts of Climate Change on the Proposed Lake Tapps–White River Water Supply*. Master's Thesis, Department of Civil and Environmental Engineering, University of Washington. (Also at http://cses.washington.edu/cig/)

Baron, J. S., N. L. Poff, P. L. Angermeier, C. N. Dahm, P. H. Gleick, N. G. Hairston, Jr., R. B. Jackson, C. A. Johnston, B. D. Richter, and A. D. Steinman. 2003. Sustaining healthy freshwater ecosystems. *Issues in Ecology* Winter: 10. http://www.esa.org/science_resources/Issues/TextIssues/issue10.php.

Barry, John M. 1997. *Rising Tide: The Great Mississippi Flood of 1927 and How It Changed America*. New York: Simon & Schuster.

Beckinsale, R. P. 1969. The Human Use of Open Channels. In *Water, Earth and Man*, ed. R. J. Charley, 331–343. London: Methuen & Co.

Bergen, S., S. Bolton, and J. Fridley. 2001. Ecological engineering: Design based on ecological principles. *Ecological Engineering* 18(2): 201–210.

Bollens, John C. 1957. *Special Districts Governments in the United States*, chapters 1 and 5. University of California Press.

Brookes, A. 1996. Floodplain restoration and rehabilitation. In *Floodplain Processes*, eds. M. G. Anderson, D. E. Walling, and P. D. Bates, 553–576. New York: Wiley.

Burby, Raymond J., and Steven P. French. 1985. *Flood Plain Land Use Management: A National Assessment*. Boulder, Colo.: Westview Press.

California Department of Water Resources Independent Review Panel. 2007. *A California Challenge—Flooding in the Central Valley*. California Department of Water Resources. http://www.water.ca.gov/news/newsreleases/2008/101507challenge.pdf.

Changnon, Stanley A., ed. 1996. *The Great Flood of 1993: Causes, Impacts, and Responses*. Boulder, Colo.: Westview Press.

Chow, V. T. 1959. *Open-Channel Hydraulics*, chapters 3 and 5. New York: McGraw-Hill.

Criss, R. E., and E. L. Shock. 2001. Flood enhancement through flood control. *Geology* October: 875–878.

Daily, G. C., S. Alexander, P. R. Ehrlich, L. Goulder, J. Lubchenco, P. A. Matson, H. A. Mooney, S. Postel, S. H. Schneider, D. Tilman, G. M. Woodwell. 1997. Ecosystem services: Benefits supplied to human societies by natural ecosystems. *Issues in Ecology* Spring: 2. http://www.esa.org/science_resources/Issues/TextIssues/issue2.php.

Deyle, R. E., S. P. French, R. B. Olshansky, and R. G. Patterson. 1998. Hazard assessment: The factual basis for planning and mitigation. In *Cooperating with Nature: Confronting Natural Hazards with Land-Use Planning for Sustainable Communities*, ed. R. J. Burby. Washington, D.C.: Joseph Henry Press.

Ellet, Charles. 1852. *Report of the Overflows of the Delta of the Mississippi River*. Washington, D.C.: A. B. Hamilton.

Federal Emergency Management Agency. 1995. *FEMA-258: Guide to Flood Insurance Rate Maps—For understanding how to read and use a FEMA Flood Insurance Rate Map*. Washington, D.C.: FEMA.

———. 2001. *FEMA 386: Understanding Your Risks. Identifying Hazards and Estimating Losses: State and Local Mitigation Planning How-To Guide*. Washington, D.C.: FEMA.

———. 2007. *Best Practices: Mitigation in Wisconsin*. Washington, D.C.: FEMA. (Produced in cooperation with the state of Wisconsin.)

Federal Interagency Stream Restoration Working Group. 1998. *Stream Corridor Restoration: Principles, Processes, and Practices*. Washington, D.C.: Government Printing Office.

Freitag, Bob, Rhonda Montgomery, and Ron Throupe. 2003. Market Impacts on Elevated Homes in a Known Floodplain: A Case Study. Presented at American Real Estate Society conference.

Freyfogle, Eric T. 2003. *The Land We Share: Private Property and the Common Good*. Washington, D.C.: Shearwater Books.

Frissell, C. A. 1997. Ecological principles. In *Watershed Restoration: Principles and Practices*, eds. J. E. Williams, C. A. Wood, and M. P. Dombeck, 96–115. Bethesda, Md.: American Fisheries Society.

Giller, P. S., and B. Mamqvist. 1998. *The Biology of Streams and Rivers*. Oxford: Oxford University Press.

Godschalk, D. R., T. Beatley, P. Berke, D. J. Brower, E. J. Kaiser, C. C. Bohl, and R. M. Goebel. 1999. *Natural Hazard Mitigation: Recasting Disaster Policy and Planning*, chapter 2. Washington, D.C.: Island Press.

Gustanski, J. A., and R. H. Squires. 2000. *Protecting the Land: Conservation Easements Past, Present, and Future*. Washington, D.C.: Island Press.

Hill, Kim Quaile, and Kenneth R. Mladenka. 1992. *Democratic Governance in American Cities and States*, chapter 3. Pacific Grove, Calif.: Brooks/Cole Publishing Company.

Hill, Libby. 2000. *The Chicago River: A Natural and Unnatural History*. Chicago: Lake Claremont Press.

Holling, C. S. 1996. Engineering versus ecological resilience. In *Engineering within Ecological Constraints*, ed. P. E. Schulze, 31–44. Washington, D.C.: National Academy of Engineering.

Hunt, Constance Elizabeth, and Verne Huser. 1988. *Down by the River: The Impact of Federal Water Projects on Biological Diversity*. Washington, D.C.: Island Press.

Johnson, Gary P., Robert R. Holmes, Jr., and Loyd A. Waite. 2003. *The Great Flood of 1993 on the Upper Mississippi River—Ten Years Later*. U.S. Geological Survey.

Junk, W. J., P. B. Bayley, and R. E. Sparks. 1989. The flood pulse concept in river-floodplain systems. *Canadian Special Publication of Fisheries and Aquatic Sciences* 106: 110–127.

Kasprisin, Ron, and James Pettinari. 1990. *Visual Thinking for Architects and Designers: Visualizing Context in Design*. Hoboken, N.J.: Wiley.

Knighton, David. 1998. *Fluvial Forms and Processes: A New Perspective*, chapter 6. New York: Oxford University Press.

Lee, Douglas. 1983. The land of the river. *National Geographic Magazine* August: 226–252.

Leopold, Luna B., M. Gordon Wolman, and John P. Miller. 1964. *Fluvial Processes in Geomorphology*. San Francisco, CA: W. H. Freeman & Co. Republished in 1995 by New York: Dover Press.

Lynch, Kevin. 1965. *The Image of the City*. Cambridge, Mass.: MIT Press.

Marsh, William M. 2005. *Landscape Planning: Environmental Applications*, 4th edition. Hoboken, N.J.: Wiley.

Mileti, Dennis S. 1999. *Disasters by Design: A Reassessment of Natural Hazards in the United States*. Washington, D.C.: Joseph Henry Press.

Montgomery, D. R. 1999. Process domains and the river continuum. *Journal of the American Water Resources Association* 36: 397–410.

Morgan, A. E. 1971. *Dams and Other Disasters: A Century of the Army Corps of Engineers in Civil Works*. Boston: Porter Sargent.

Naeem, S., F. S. Chapin III, R. Costanza, P. R. Ehrlich, F. B. Golley, D. U. Hooper, J. H. Lawton, R. V. O'Neill, H. A. Mooney, O. E. Sala, A. J. Symstad, and D. Tilman. 1999. Biodiversity and ecosystem functioning: Maintaining natural life support processes. *Issues in Ecology* Fall: 4. http://www.epa.gov/watertrain/pdf/issue4.pdf.

National Institute of Building Sciences, The Multihazard Mitigation Council. 2005. *Natural Hazard Mitigation Saves: An Independent Study to Assess the Future Savings from Mitigation Activities*. Washington, D.C.: National Institute of Building Sciences.

National Research Council. 1999. *New Strategies for America's Watersheds*. Washington, D.C.: National Academy Press.

North Carolina Hazard Mitigation Section. April 2000. *Getting to Open Space: Alternatives to Demolition and Options for Land Use: A Guide for Hazard Mitigation Grant Program Acquisition Projects*. North Carolina Department of Crime Control and Public Safety, Division of Emergency Management.

Pearce, Fred. 2006. *When the Rivers Run Dry: Water—The Defining Crisis of the Twenty-first Century*. Boston: Beacon.

Ricklefs, R. E. 1990. *Ecology*. New York: W. H. Freeman.

Riley, Ann. 1998. *Restoring Streams in Cities: A Guide for Planners, Policymakers, and Citizens*. Washington, D.C.: Island Press.

Schwab, Jim. 1996. "Nature bats last": The politics of floodplain management, environment, and development. In the January/February newsletter of the American Planning Association.

Shepp, D. L., and J. D. Cummings. 1997. Restoration in an urban watershed: Anacostia River of Maryland and the District of Columbia. In *Watershed Restoration: Principles and Practices*, eds. J. E. Williams, C. A. Wood, and M. P. Dombeck, 297–317. Bethesda, Md.: American Fisheries Society.

Skinner, B. J., and S. C. Porter. 2000. *The Dynamic Earth*, 4th edition. Hoboken, N.J.: Wiley.

Sverdrup, K. A., A. B. Duxbury, and A. C. Duxbury. 2003. *An Introduction to the World's Oceans*, 7th edition. New York: McGraw-Hill.

Task Force on the Natural and Beneficial Function of the Floodplain. 2002. *FEMA 409: The Natural and Beneficial Functions of Floodplains: Reducing Flood Losses by Protecting and Restoring the Floodplain Environment*. A Report for Congress. FEMA.

Thomas, Edward A. 2006. Overcoming legal challenges: A perfect storm of opportunities. *Natural Hazards Observer* XXXI(2): 11–12.

Tockner, K., F. Malard, and J. V. Ward. 2000. An extension of the flood pulse concept. *Hydrological Processes* 14: 2861–2883.

Waldner, Leora S. 2007. Floodplain creep and beyond: An assessment of next-generation floodplains problems. *Journal of Emergency Management* July/August: 39–46.

Ward, J. V. 2001. Biodiversity: Towards a unifying theme for river ecology. *Freshwater Biology* 46: 807–819.

Watt, James G. 1983. Economic and Environmental Principles and Guidelines for Water and Related Land Resources Implementation Studies. *US Water Resources Council*.

White, G. F., ed. 1974. *Natural Hazards: Local, National, Global*, chapter 1. New York: Oxford University Press.

Williams, J. E., C. A. Wood, and M. P. Dombeck. 1997. Understanding watershed-scale restoration. In *Watershed Restoration: Principles and Practices*, eds. J. E. Williams, C. A. Wood, and M. P. Dombeck, 1–13. Bethesda, Md.: American Fisheries Society.

Worster, Donald. 1985 (1992). *Rivers of Empire: Water, Aridity, and the Growth of the American West*. New York: Oxford University Press.

On the Internet

Association of State Floodplain Managers. 2004. *No adverse impact white paper*. http://www.floods.org.

Association of State Floodplain Managers Foundation. 2004a. *A collection of papers prepared for the September, 2004, National Policy Forum: Reducing flood losses: Is the 1% chance (100-year) flood standard sufficient?* http://www.floods.org.

———. 2004b. *Floodplain management: State and local programs 2003*. http://www.floods.org.

Bolton, S. M., and J. Shellberg. *Ecological issues in floodplain and riparian corridors: White Paper for the state of Washington.* http://wdfw.wa.gov/hab/ahg/floodrip.htm.

Brower, David. Session 9: Hazard mitigation planning. In *Principles and Practice of Hazard Mitigation,* FEMA Higher Education Course. http://training.fema.gov/emiweb/edu/completeCourses.asp.

Campbell, Warren. 2007. Western Kentucky University Stormwater Utility Survey. http://www.wku.edu/swusurvey/SWU%20Survey%202008.pdf.

Cincinnati. http://www.cincinnatiport.org and http://www.crpark.org.

Clark, J. L. 1999. Effects of Urbanization on Streamflow in Three Basins in the Pacific Northwest. Master's Thesis, Portland State University. http://nwdata.geol.pdx.edu/Thesis/FullText/1999/Clark/.

Federal Emergency Management Agency How-To Guides:

Getting Started. http://www.fema.gov/library/viewRecord.do?id=1867.

Understanding Your Risks. http://www.fema.gov/library/viewRecord.do?id=1880.

Developing the Mitigation Plan. http://www.fema.gov/library/viewRecord.do?id=1886.

Bringing the Plan to Life. http://www.fema.gov/library/viewRecord.do?id=1887.

Using Benefit-Cost Review. http://www.fema.gov/library/viewRecord.do?id=2680.

Integrating Historic Property. http://www.fema.gov/library/viewRecord.do?id=1892.

Integrating Manmade Hazards. http://www.fema.gov/library/viewRecord.do?id=1915.

Multi-Jurisdictional Mitigation Planning. http://www.fema.gov/library/viewRecord.do?id=1905.

Federal Emergency Management Agency. *FEMA 480: Floodplain Management Requirements: A study guide and desk reference for local officials.* http://www.fema.gov/plan/prevent/floodplain/fm_sg.shtm.

———. 2003. *35 Years of NFIP Highlights.* Number 2. 30–32. Washington, D.C. http://www.fema.gov/pdf/nfip/wm2003_2.pdf.

Friends of the River. *Beyond flood control: Flood management and river restoration.* http://www.friendsoftheriver.org/fotr/BeyondFloodControl/fldprimr.html.

Gertz, Emily. 2008. Tempting fate: Fifteen years after the Great Flood of 1993, floodplain development is booming. In *Grist.* http://www.grist.org/feature/2008/03/19/gertz/.

Healy, Andrew J., and Neil Malhotra. 2008. *Preferring a pound of cure to an ounce of prevention: Retrospective voting and failures in electoral accountability.* http://www.sscnet.ucla.edu/polisci/cpworkshop/papers/Healy.pdf.

Iowa Natural Heritage Foundation. *Lessons from the last big flood.* http://www.inhf.org/mag-2008summer/magsum08-flood-lessons.htm.

Kious, W. J., and R. I. Tilling. 1996. *This dynamic Earth: The story of plate tectonics.* U.S. Geological Survey. http://pubs.usgs.gov/publications/text/dynamic.html.

Larson, L., and D. Plasencia. 2004. *No adverse impact floodplain management.* Published in *Natural Hazards Review,* November 2001. IAAN 1527-6988. http://www.floods.org.

Morrison, Jim. 2005. How Much Is Clean Water Worth? *National Wildlife Magazine* 43(2). http://www.nwf.org/NationalWildlife/article.cfm?issueID=73&articleID=1032.

National Cave and Karst Research Institute and U.S. Geological Survey. *The national karst map.* http://www.nature.nps.gov/nckri/map/project/index.html.

National Environmental Policy Act Process (40 CFR 1500–508). http://www.epa.gov/compliance/basics/nepa.html.

National Marine Fisheries Service. 2008. *Endangered Species Act—Section 7 Consultation Final Biological Opinion and Magnuson-Stevens Fishery Conservation and Management Act Essential Fish Habitat Consultation: Implementation of the National Flood Insurance Program in the State of Washington Phase One Document—Puget Sound Region*. http://pcts.nmfs.noaa.gov/pls/pcts-pub/pcts_upload.summary_list_biop?p_id=29082.

Natural Resources Conservation Service. 2001. *Stream corridor restoration*. http://www.nrcs.usda.gov/technical/stream_restoration/newtofc.htm.

National Weather Service. 1992. *Flash floods and floods . . . the awesome power!* http://www.weather.gov/om/brochures/ffbro.htm.

O'Keefe, T., R. Naiman, S. Eliot, and D. Norton. *Introduction to watershed ecology*. Environmental Protection Agency, Office of Wetlands, Oceans and Watersheds. http://www.epa.gov/watertrain/ecology/index.html.

Pidwirny, M. 2004. *Fundamentals of physical geography*. http://www.physicalgeography.net/home.html.

Seattle Post-Intelligencer. December 10, 2008. Timber, but no homes, on 45,500-acre swath. *Seattle Post-Intelligencer*. http://www.seattlepi.com/local/391582_plumcreek11.html.

Sisk, T. D., ed. 1999. *Perspectives on the land use history of North America: A context for understanding our changing environment*. USGS Biological Resources Division, Science Report USGS/BRD/BSR-1998-000 (revised September 1999). http://biology.usgs.gov/luhna/

Tulsa. *Flood control and drainage*. http://www.cityoftulsa.org/CityServices/FloodControl/History.asp.

U.S. Department of Energy. 1994. *Rebuilding for the Future: A Guide to Sustainable Redevelopment for Disaster-Affected Communities*. http://www.smartcommunities.ncat.org/articles/RFTF1.shtml.

U.S. Fish and Wildlife Service. *Horseshoe Bend Division*. http://www.fws.gov/Midwest/PortLouisa/horseshoe_bend.html.

Index

active terraces: floodplain creation and, 55–56; polygons and, 216, *217*

agriculture: easements and, 15–16, 17; flood damage and, 15; floodplains and, 85–86; levees and, 15, 16, 17; river history and, 75–76, 77–79; rivers and, 75–76, 77–79, 85–86

analysis: benefit-cost, 162, 164–165; for flood mitigation projects, 190; watershed, 9. *See also* strategy decision-making tools

animals, 60–61, 84

approaches, 93, 190–191; Buck Hollow River, Oregon case study, 109–112; four, 94; graphic tool kit and, 108–109; historical, 69; land use, 104–107; not addressed by NFIP, 205; risks/tools and, 139–140; strategies/tools and, 137; urban riverfronts and urban design, 107–108; to watersheds, 101, *102*, 109–112. *See also* nonstructural approaches; structural approaches

Arkansas River, 182, 183

assets: floodplain, 84–85; points as watershed, 215, 219–220, *221*; protection and flood control, 1; river, 84–85. *See also* benefits; floodplain asset/value enhancement/protection; values

Association of State Floodplain Managers, 19, 30, 36

base flood elevation (BFE), 207

basins: nonstructural approaches and river, 102; structural approaches and detention, 94. *See also* sub-basins

benefits: -cost analysis, 162, 164–165; definition of, 21; flood management for, 19;

of floodplains, 73–74; of floods, 19, 21, 22, 23, 41; opportunity and, 22; of rivers, 73–74; of watersheds, 73–74. *See also* assets; values

BFE. *See* base flood elevation

biological change: floodplains and, 82–84; human activity and, 82–84; polygons and, 216, *218*; rivers and, 82–84; values and, 138–139

Buck Hollow River, Oregon, 109–112

Bureau of Reclamation, 119, 131

California Central Valley, 100–101

capabilities: Davenport, Iowa case study, 133–135; definition of, 21, 113; floods and, 21, 22; integrating tools and, 132–133; money and, 113–114; opportunity and, 22, 24, *25*; political power as, 113, 114; risks and, 21, 22, 24, *25*; strategy assessment of, 173, *176–177*, 178; timing and, 113, 114–118

Carter, Jimmy, 183

case studies: Buck Hollow River, Oregon, 109–112; Chicago, Illinois, 89–91; Davenport, Iowa, 133–135; Fife, Washington, 209–214; Louisa County, Iowa, 14–17; New York, New York, 69–72; NFIP and, 36, 37, 38, 39; Rivergrove, 194–195; Snoqualmie, Washington, 36–39; Soldiers Grove, Wisconsin, 47–51; Tulsa, Oklahoma, 182–187; USACE and, 48, 90, 134; Washington I-5 flooding, 153–158

Centralia Flood Damage Reduction Project, 155–156

CEQ. *See* Council on Environmental Quality